LECTURES ON RIEMANN SURFACES,

JACOBI VARIETIES

BY

R. C. GUNNING

PRINCETON UNIVERSITY PRESS

AND

UNIVERSITY OF TOKYO PRESS

PRINCETON, NEW JERSEY
1972

L.C. Card: 72-9950

I.S.B.N. 0-691-08127-1

Published in Japan exclusively
by the University of Tokyo Press;
in other parts of the world by
Princeton University Press

Printed in the United States of America

Preface.

These are notes based on a course of lectures given at Princeton University during the Spring Term of 1972. They are intended as a sequel to the notes Lectures on Riemann Surfaces (Mathematical Notes, Princeton University Press, 1966), and are principally devoted to a continuation of the discussion of the dimensions of spaces of holomorphic cross-sections of complex line bundles over compact Riemann surfaces contained in §7 of the earlier notes; but whereas the earlier treatment was limited to results obtainable chiefly by one dimensional methods, a more detailed analysis requires the use of various properties of Jacobi varieties and of symmetric products of Riemann surfaces, and consequently serves as a further introduction to these topics as well.

The reader is assumed to be familiar with the material covered in the first 8 chapters of the earlier notes, and the terminology and notation introduced there are used with no further explanation. The first chapter of these notes consists of the rather explicit description of a canonical basis for the Abelian differentials on a marked Riemann surface, and of the description of the canonical meromorphic differentials and the prime function (in more or less the sense used in Hensel and Landsberg "Theorie der algebraische Funktionen einer Variabeln") of a marked Riemann surface. Although most of this material is used only to a minor extent in the present volume of notes, it is important for later developments and this is a convenient point at which it can be treated, particularly in connection with the introduction of the very useful concept of a marked Riemann surface, which concept is employed systematically throughout the remainder of these notes. The second chapter contains a discussion of Jacobi varieties of compact Riemann surfaces and of various subvarieties which arise in determining the dimensions of spaces of holomorphic cross-sections of complex line bundles. The Jacobi variety of a marked Riemann surface is introduced in an explicit but convenient manner, and the usual invariance properties follow easily from an identification of the Picard and Jacobi varieties. The third chapter contains a discussion of the relations between Jacobi varieties and symmetric products of Riemann surfaces which are relevant to the determination of dimensions of spaces of holomorphic cross-sections of complex line bundles. The analytical and topological properties of symmetric products are not really discussed in general, although an indication of the usefulness of some of these properties in studying the problem at hand is indicated in the appendix. The final chapter consists of the derivation of Torelli's theorem following A. Weil but in an analytic context.

The results which may be new are somewhat scattered and follow fairly naturally from the lines of development of the subject, so there is no purpose served in attempting to point them out explicitly. However I have tried in the notes and references to indicate rather explicitly the sources used for the material covered here, but without attempting a complete bibliography or detailed and accurate history of the subject. As is the case with so many other publications on this subject the debts owed to the work of A. Weil are immense and obvious. I have also been heavily influenced by the work of H. H. Martens and have followed his treatment at several places indicated in the notes. Finally I should like to express my thanks to the students and colleagues who attended these lectures, particularly to Robert H. Risch, for many helpful suggestions and references; and to Elizabeth Epstein for her customary beautiful job of typing these notes.

R. C. Gunning

Princeton, New Jersey
September, 1972

Contents

§1. Marked Riemann surfaces and their canonical differentials.

(a) At several points in the more detailed study of Riemann surfaces, the explicit topological properties of surfaces play an important role; and it is convenient to have these properties established from the beginning of the discussion, to avoid the necessity of inserting topological digressions later. Since the universal covering space of a connected orientable surface of genus $g > 0$ is a cell, the fundamental group carries essentially all the topological properties of the surface; so it is also convenient to introduce from the beginning and to use systematically henceforth the representation of a Riemann surface in terms of its universal covering space.

Let M be a compact Riemann surface of genus $g > 0$, and let $\tilde{M}$ be its universal covering space. The topological space $\tilde{M}$ inherits from M in an obvious manner a complex analytic structure, hence $\tilde{M}$ is itself a Riemann surface; and $\tilde{M}$ is topologically quite trivial, being homeomorphic to an open disc. The covering transformations form a group Γ of complex analytic homeomorphisms $T: \tilde{M} \longrightarrow \tilde{M}$; and the Riemann surface M can be identified with the quotient space $\tilde{M}/\Gamma$. It will be assumed that the reader is familiar with the topological properties of covering spaces, so that no further details need be given here.

Select a base point $p_o \in M$ and also a base point $z_o \in \tilde{M}$ lying over p_o. Having made these selections, there is a canonical isomorphism between the covering transformation group Γ and the fundamental group $\pi_1(M,p_o)$ of the surface M based at p_o; this is the isomorphism which associates to any transformation $T \in \Gamma$ the class of loops in $\pi_1(M,p_o)$ represented by the image in M of any path from z_o to Tz_o in $\tilde{M}$. Again the details will be omitted, since they can be supplied in a quite straightforward manner by anyone familiar with the topological properties of covering spaces; but it should be noted that this isomorphism does depend on the choice of the base point z_o, since the selection of another base point in $\tilde{M}$ lying over p_o alters the isomorphism by an inner automorphism of the group $\pi_1(M,p_o)$.

Now for the more detailed properties, recall that topologically M is just a sphere with g handles; hence M can be dissected into a connected contractible set by cutting along 2g paths, as indicated in the accompanying diagram.

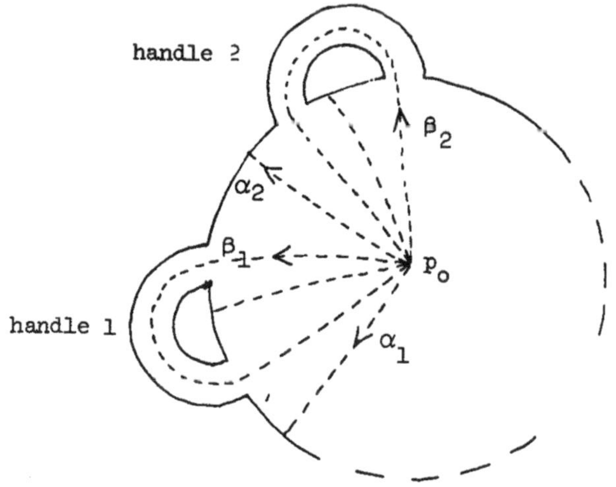

Each loop α_i or β_i lifts to a unique path $\tilde{\alpha}_i$ or $\tilde{\beta}_i$ in $\tilde{M}$ beginning at the base point z_o ; the path $\tilde{\alpha}_i$ runs from z_o to $A_i z_o$, where $A_i \in \Gamma$ is the transformation associated to the homotopy class of α_i in $\pi_1(M,p_o)$ under the isomorphism introduced above, and the path $\tilde{\beta}_i$ runs from z_o to $B_i z_o$, where $B_i \in \Gamma$ is associated to β_i in the corresponding manner. The complement $M - \bigcup_{i=1}^{g}(\alpha_i \cup \beta_i)$ is simply connected, hence lifts homeomorphically to a number of disjoint open subsets of $\tilde{M}$ which are permuted by the action of any element of the covering transformation group. It follows readily, upon tracing out the boundary of this complement in the preceding diagram, that one of these liftings has the form indicated in the following diagram.

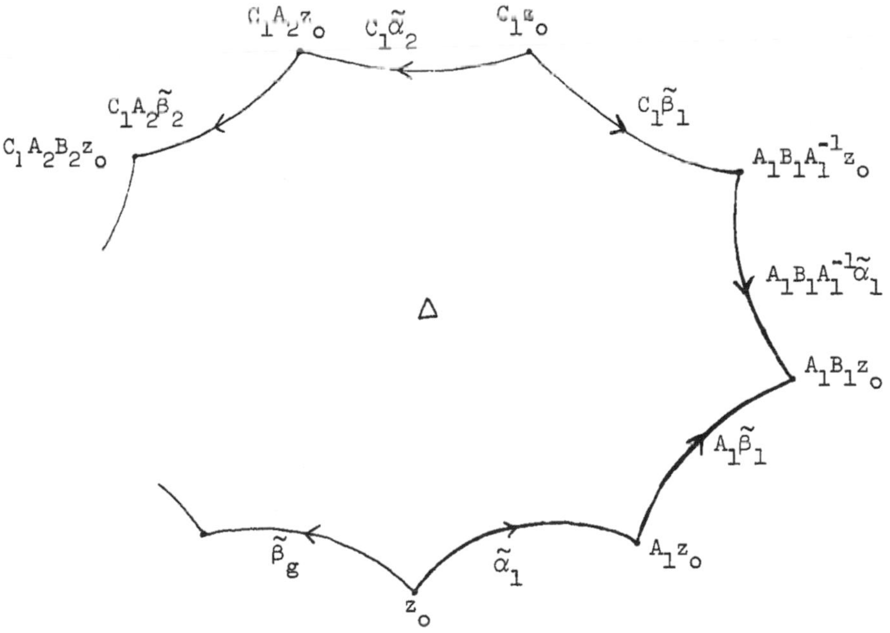

In this diagram the elements $C_i \in \Gamma$ are the commutators

(1) $$C_i = A_i\, B_i\, A_i^{-1}\, B_i^{-1} \qquad (i = 1,\ldots,g) ;$$

and the domain $\Delta \subset \tilde{M}$ as indicated is homeomorphic to the complement $M - \bigcup_{i=1}^{g} (\alpha_i \cup \beta_i)$. The point set closure $\bar{\Delta}$ of this domain is a polygonally shaped subset of $\tilde{M}$ with boundary consisting of the closed curve

(2) $$\partial\Delta = \prod_{i=1}^{g} (C_1\cdots C_{i-1}\tilde{\alpha}_i)\cdot(C_1\cdots C_{i-1}A_i\tilde{\beta}_i)\cdot(C_1\cdots C_iB_i\tilde{\alpha}_i)^{-1}\cdot(C_1\cdots C_i\tilde{\beta}_i)^{-1}.$$

As usual in the discussion of the groupoid of paths, a product $\tilde{\alpha}\cdot\tilde{\beta}$ of paths is the path obtained by traversing first $\tilde{\alpha}$ and then $\tilde{\beta}$; this multiplication is noncommutative, but is defined only when the end point of $\tilde{\alpha}$ coincides with the initial point of $\tilde{\beta}$, hence the notation used in (2) should lead to no confusion. As is well known, the group Γ is generated by the 2g elements $A_1,\ldots,A_g,B_1,\ldots,B_g$; and these generators are subject to the single relation

(3) $$C_1C_2 \cdots C_g = 1 ,$$

where again C_i denotes the commutator (1). The fundamental group $\pi_1(M,p_o)$ is of course correspondingly generated by the loops $\alpha_1,\ldots,\alpha_g,\beta_1,\ldots,\beta_g$, subject to the corresponding relation; that this relation does hold is obvious from (2), although it does require some more work to show that all other relations are consequences of (3).

A selection of base points $p_o \in M$, $z_o \in \tilde{M}$ and a set of cuts $\alpha_1,\ldots,\alpha_g,\beta_1,\ldots,\beta_g$ of the above canonical form will be called a <u>marking</u> of the Riemann surface M. A marked Riemann surface thus has a specified base point, a fixed isomorphism between the covering transformation group of its universal covering space and the fundamental group at its chosen base point, a canonical set of generators for that covering transformation group and hence for the fundamental group, and a canonical dissection of the universal covering space into polygonally shaped subsets $\Gamma\Delta = \{T\Delta | T \in \Gamma\}$. The set Δ will be called the standard fundamental domain for the action of the covering transformation group Γ on the universal covering space $\tilde{M}$; and the notation introduced above for the canonical generators of Γ will be used consistently in the sequel.

There are of course a vast number of possible markings for any given Riemann surface. As noted above, the choice of another base point in $\tilde{M}$ has merely the effect of altering the isomorphism between Γ and $\pi_1(M,p_o)$ by an inner automorphism, hence of altering the canonical generators of Γ by an inner automorphism of Γ while of course leaving the canonical generators of $\pi_1(M,p_o)$ unchanged, and of replacing the standard fundamental domain Δ by a suitable translate of Δ under the action of Γ. An arbitrary orientation preserving homeomorphism of the topological space M clearly transforms any marking into another marking, modulo choices of base points in $\tilde{M}$; and it is easy to see that conversely any two markings of M can be transformed into one another by some orientation preserving homeomorphism of the underlying topological space M , again modulo choices of base points in $\tilde{M}$, since the corresponding standard fundamental domains in $\tilde{M}$ are evidently homeomorphic under an orientation preserving homeomorphism commuting with Γ . Thus, holding the base points $p_0 \in M$, $z_0 \in \tilde{M}$ fixed for simplicity, all possible markings arise from a given marking by applying suitable orientation preserving homeomorphisms of M to itself leaving p_0 fixed. Note that any such homeomorphism determines in turn an automorphism of the fundamental group $\pi_1(M,p_0)$, taking the canonical set of generators corresponding to one marking into the canonical set of generators corresponding to another marking. A homeomorphism which is homotopic to the identity (through homeomorphisms leaving the base point fixed) clearly yields the identity automorphism of the fundamental group, so that the canonical sets of generators of the fundamental group $\pi_1(M,p_0)$ associated to two markings so related actually coincide; two markings so related will be called equivalent, and the reader should be warned that in much of the literature only these equivalence classes of markings are really considered. The more detailed investigation of these questions is an interesting subject in its own right, but must be left aside at present.

The paths $\alpha_1,\dots,\alpha_g,\beta_1,\dots,\beta_g$ can be viewed as singular cycles on M , and as such represent a basis for the singular homology group $H_1(M,\mathbf{Z})$; thus a marked Riemann surface also has a canonical set of generators for its first homology group. The intersection properties of these one-cycles can be read off immediately from the diagram on page 2. Note that the choice of another marking on the surface, by the application of a homeomorphism of M to itself, determines an automorphism of the homology group $H_1(M,\mathbb{Z})$; this automorphism can also be determined directly from the corresponding automorphism of $\pi_1(M,\mathbb{Z})$, recalling that $H_1(M,\mathbb{Z})$ is just the abelianization of $\pi_1(M,\mathbb{Z})$. It should be mentioned that there are nontrivial automorphisms of $\pi_1(M,\mathbb{Z})$ which induce trivial automorphisms of $H_1(M,\mathbb{Z})$; there is a real and important distinction between properties of the surface which depend on homological and those which depend on homotopical properties, as will become evident later.

(b) The Abelian differentials on M are the holomorphic differential forms of type $(1,0)$; they form a g-dimensional complex vector space $\Gamma(M, \mathcal{O}^{1,0})$. Note that any Abelian differential $\omega \in \Gamma(M, \mathcal{O}^{1,0})$ can be viewed as a Γ-invariant holomorphic differential form of type $(1,0)$ on the universal covering surface $\tilde{M}$; this form will also be denoted by ω , a notational convention that really leads to no confusion since M can be identified with the quotient space $\tilde{M}/\Gamma$. Since $\tilde{M}$ is simply connected and since any such Abelian differential ω is closed, there must exist a holomorphic function w on $\tilde{M}$ such that $\omega = dw$; such a function is called an Abelian integral for the Riemann surface. Note that the function w is determined uniquely up to an additive constant.

For a Riemann surface with a specified base point $z_o \in \tilde{M}$, the associated Abelian integral can be normalized so that $w(z_o) = 0$, and can thus be viewed as determined uniquely by the Abelian differential ω; indeed the Abelian integral is then given explicitly by the integral

$$w(z) = \int_{z_o}^{z} \omega .$$

Note further that since the Abelian differential ω is Γ-invariant, the Abelian integral $w(z)$ has the property that $d[w(Tz) - w(z)] = \omega(Tz) - \omega(z) = 0$ for any $T \in \Gamma$, hence that $w(Tz) = w(z) - \omega(T)$ for some constant $\omega(T) \in \mathbf{C}$ depending on $T \in \Gamma$. It then follows readily that $\omega(ST) = \omega(S) + \omega(T)$ for any two elements $S, T \in \Gamma$; hence the set of these constants can be viewed as an element $\omega \in \mathrm{Hom}(\Gamma, \mathbf{C})$, which will be called the <u>period class</u> of the Abelian differential $\omega \in \Gamma(M, \mathcal{O}^{1,0})$. This terminology is suggested by the observation that $\omega(T) = -\int_{z_o}^{Tz_o} \omega$ for any $T \in \Gamma$. Note finally that an Abelian differential is determined uniquely by its period class; for if $\omega_1, \omega_2 \in \Gamma(M, \mathcal{O}^{1,0})$ have the same period class, their difference $\omega_1 - \omega_2 = d[w_1 - w_2]$ is the derivative of a holomorphic Γ-invariant function on $\tilde{M}$, hence that difference must vanish since the only holomorphic functions on the compact Riemann surface $\tilde{M}/\Gamma$ are constants.

Now select a marking for the Riemann surface M and a basis $\omega_1, \ldots, \omega_g$ for the Abelian differentials on M. The period classes of these Abelian differentials are determined by their values $\omega_i(A_j)$, $\omega_i(B_j)$ on the canonical generators for the covering transformation group Γ ; these values can be grouped together to form the associated $g \times 2g$ period matrix (Ω', Ω'') , where $\Omega' = \{\omega_i(A_j)\}$, $\Omega'' = \{\omega_i(B_j)\}$ are $g \times g$ matrices.

Theorem 1. The period matrix (Ω', Ω'') of a basis for the Abelian differentials on a marked Riemann surface M satisfies the conditions

(i) $\Omega' \cdot {}^t\Omega'' - \Omega'' \cdot {}^t\Omega' = 0$, (Riemann's equality), and

(ii) $i\Omega' \cdot {}^t\overline{\Omega}'' - i\Omega'' \cdot {}^t\overline{\Omega}'$ is positive definite Hermitian, (Riemann's inequality).

Proof. Although this was proved in the earlier lecture notes (Theorem 17), it is perhaps worthwhile repeating that proof to show how the intersection matrix of the canonical basis for the one-cycles on a marked Riemann surface can be calculated and to serve as a model for several quite similar later calculations. The essential point in deriving Riemann's equality is that $\omega_i \wedge \omega_j = 0$; viewing these as forms on $\widetilde{M}$ and integrating over the standard fundamental domain Δ , and recalling that the boundary of Δ has the form given in (2), it follows that:

$$0 = \int_{\Delta} \omega_i \wedge \omega_j = \int_{\Delta} d(w_i \omega_j) = \int_{\partial\Delta} w_i \omega_j$$

$$= \sum_{k=1}^{g} \left\{ \int_{\tilde{\alpha}_k} w_i(C_1 \cdots C_{k-1} z) \quad \omega_j(C_1 \cdots C_{k-1} z) \right.$$

$$+ \int_{\tilde{\beta}_k} w_i(C_1 \cdots C_{k-1} A_k z) \quad \omega_j(C_1 \cdots C_{k-1} A_k z)$$

$$- \int_{\tilde{\alpha}_k} w_i(C_1 \cdots C_k B_k z) \quad \omega_j(C_1 \cdots C_k B_k z)$$

$$\left. - \int_{\tilde{\beta}_k} w_i(C_1 \cdots C_k z) \quad \omega_j(C_1 \cdots C_k z) \right\}$$

$$= \sum_{k=1}^{g} \left\{ \int_{\tilde{\alpha}_k} w_i(z)\omega_j(z) + \int_{\tilde{\beta}_k} [w_i(z) - \omega_i(A_k)]\omega_j(z) \right.$$

$$\left. - \int_{\tilde{\alpha}_k} [w_i(z) - \omega_i(B_k)]\omega_j(z) - \int_{\tilde{\beta}_k} w_i(z)\omega_j(z) \right\}$$

$$= \sum_{k=1}^{g} \left\{ \omega_i(A_k)\omega_j(B_k) - \omega_i(B_k)\omega_j(A_k) \right\} ,$$

which establishes (1). The essential point in deriving Riemann's inequality is that for any Abelian differential $\omega(z) = f(z)dz$ in a local coordinate system $z = x + iy$

$$i\omega(z)\wedge\overline{\omega}(z) = |f(z)|^2 \, idz\wedge d\overline{z} = 2|f(z)|^2 \, dx\wedge dy ,$$

hence that

$$\int_{\Delta} i\omega(z)\wedge\overline{\omega}(z) \geqq 0$$

with equality holding only when $\omega = 0$. In particular, putting $\omega(z) = \sum_{i=1}^{g} t_i\omega_i(z)$ for arbitrary complex constants t_i, it follows that

$$\sum_{i,j=1}^{g} P_{ij}t_i\bar{t}_j \geqq 0$$

with equality holding only when $t_1 = \dots = t_g = 0$, where

$$P_{ij} = \int_{\triangle} i\omega_i(z)\wedge\bar{\omega}_j(z) \;;$$

thus the matrix $P = \{P_{ij}\}$ is positive definite Hermitian. To determine this matrix explicitly, it follows as above that

$$-iP_{ij} = \int_{\triangle} \omega_i(z)\wedge\bar{\omega}_j(z) = \int_{\triangle} d[w_i(z)\bar{\omega}_j(z)] = \int_{\partial\triangle} w_i(z)\bar{\omega}_j(z)$$

$$= \sum_{k=1}^{g} \{\omega_i(A_k)\bar{\omega}_j(B_k) - \omega_i(B_k)\bar{\omega}_j(A_k)\} ,$$

which establishes (ii) and concludes the proof.

Consider then another basis $\tilde{\omega}_1,\dots,\tilde{\omega}_g$ for the Abelian differentials on M, where $\tilde{\omega}_i = \sum_{j=1}^{g} c_{ij}\omega_j$ for some nonsingular complex matrix $C = \{c_{ij}\}$. The period matrix associated to this basis evidently has the form $(\tilde{\Omega}',\tilde{\Omega}'') = C\cdot(\Omega',\Omega'')$. Note that an immediate consequence of Riemann's inequality is that the matrices Ω' and Ω'' are nonsingular; for if Ω' were singular it would be possible to choose a nonzero row vector $v \in \mathbb{C}^g$ such that $v\cdot\Omega' = 0$ hence such that $v\cdot(\Omega'\cdot{}^t\bar{\Omega}'' - \Omega''\cdot{}^t\bar{\Omega}')\cdot{}^t\bar{v} = 0$, contradicting Riemann's inequality, and similarly of course for Ω''. Thus there is a unique basis $\omega_1, \dots, \omega_g$ for the Abelian differentials on M such that the associated period matrix has the form (I, Ω), where I is the identity matrix; this will be called the canonical basis for the Abelian differentials on M associated to the given marking. For such a basis the period matrix is of course determined by the second component $\Omega = \{\omega_{ij}\}$, where $\omega_{ij} = \omega_i(B_j)$, while $\omega_i(A_j) = \delta^i_j$ in terms of the Kronecker symbol. In this case, Riemann's equality reduces to the assertion that the matrix Ω is symmetric, and Riemann's inequality reduces to the assertion that the imaginary part of the matrix Ω is positive definite. The matrix Ω itself will be called the canonical period matrix for the marked Riemann surface.

The canonical period matrix for a marked Riemann surface does depend on the choice of the marking; a more detailed discussion of this will be left aside at present, but it is at least worth pointing out here that the canonical period matrix is only a homological invariant. If two markings of the surface determine the same canonical generators for $H_1(M, \mathbb{Z})$, then the canonical period matrices associated to these two markings obviously coincide; thus the only analytic invariants of the surface M that can possibly be expressed as functions of the canonical period matrix Ω are those which are also homological invariants, in the same sense.

In the proof of Theorem 1, Riemann's equality was demonstrated as a direct consequence of the equalities $\omega_i \wedge \omega_j = 0$ for any pair of Abelian differentials; but a more careful examination of the proof shows that Riemann's equality is really rather weaker than the equalities $\omega_i \wedge \omega_j = 0$, indeed is equivalent to the equalities $\int_\Delta \omega_i \wedge \omega_j = 0$. That is to say, Riemann's equality merely reflects the fact that the differential forms $\omega_i \wedge \omega_j$ are homologous to zero, and not the stronger assertion that these forms actually vanish. This suggests looking more closely into this situation, to see what further properties can be deduced from the equalities $\omega_i \wedge \omega_j = 0$.

Choosing then a canonical basis $\omega_1, \ldots, \omega_g$ for the Abelian differentials on the marked Riemann surface M, introduce the holomorphic differential forms $\sigma_{ij} = w_i \omega_j \in \Gamma(\tilde{M}, \mathcal{O}^{1,0})$, where w_i are the normalized Abelian integrals. The equalities $\omega_i \wedge \omega_j = 0$ are of course precisely equivalent to the conditions that these forms σ_{ij} be closed, since $d\sigma_{ij} = \omega_i \wedge \omega_j$; and since the universal covering space $\tilde{M}$ is simply connected, these conditions are in turn equivalent to the existence of holomorphic functions $s_{ij} \in \Gamma(\tilde{M}, \mathcal{O})$ such that $\sigma_{ij} = ds_{ij}$. The functions s_{ij} are of course only determined up to an additive constant, so can be fixed uniquely by the normalization $s_{ij}(z_o) = 0$; and this normalization will be adopted systematically henceforth. Note that for any element $T \in \Gamma$ the differential forms σ_{ij} satisfy

(4) $$\sigma_{ij}(Tz) = \sigma_{ij}(z) - \omega_i(T)\omega_j(z) ,$$

hence the functions s_{ij} satisfy

(5) $$s_{ij}(Tz) = s_{ij}(z) - \omega_i(T)w_j(z) - \sigma_{ij}(T)$$

for some complex constants $\sigma_{ij}(T)$ depending on $T \in \Gamma$; the set of these constants can be viewed as a mapping $\sigma_{ij}: \Gamma \longrightarrow \mathbf{C}$, and these mappings will be called the quadratic period classes of the canonical basis for the Abelian differentials on the marked Riemann surface.

It follows immediately from (5) that these quadratic period classes satisfy the formal algebraic condition

(6) $$\sigma_{ij}(ST) = \sigma_{ij}(S) + \sigma_{ij}(T) - \omega_i(S)\omega_j(T)$$

for any elements $S,T \in \Gamma$; this is the analogue for the quadratic period classes of the condition that the ordinary period classes of Abelian differentials are homomorphisms from Γ into $\mathbf{C}$. Note that as a consequence of (6),

(7) $$\sigma_{ij}(STS^{-1}T^{-1}) = - \omega_i(S)\omega_j(T) + \omega_i(T)\omega_j(S)$$

for any commutator $STS^{-1}T^{-1} \in \Gamma$; and since the canonical generators A_i, B_i for Γ satisfy $C_1C_2 \ldots C_g = 1$ where $C_i = A_iB_iA_i^{-1}B_i^{-1}$, it follows further that

$$0 = \sigma_{ij}(C_1C_2\ldots C_g) = \sum_{k=1}^{g} (-\omega_i(A_k)\omega_j(B_k) + \omega_i(B_k)\omega_j(A_k)) .$$

Thus Riemann's equality is a formal consequence of the existence of quadratic period classes satisfying (6), reflecting the observation made above that Riemann's equality is weaker than the equations involved in the construction of the quadratic period classes. A more detailed discussion of these quadratic period classes must be left until the later discussion of general multiplicative properties.

(c) The meromorphic Abelian differentials on M are the meromorphic differential forms of type (1,0), which compose a vector space $\Gamma(M, \mathcal{M}^{1,0})$. These forms are of course determined by their singularities, up to additive holomorphic Abelian differentials; and there arises almost immediately the problem of deciding what these singularities can be. Viewing the sheaf $\mathcal{O}^{1,0}$ of germs of holomorphic differential forms of type (1,0) as a subsheaf of the sheaf $\mathcal{M}^{1,0}$ of germs of meromorphic differential forms of type (1,0), the natural quotient sheaf $\mathcal{P}^{1,0} = \mathcal{M}^{1,0}/\mathcal{O}^{1,0}$ describes these singularities locally; this will be called the sheaf of germs of principal parts of meromorphic differential forms of type (1,0). In terms of a local coordinate z in a coordinate neighborhood of M, any meromorphic differential form of type (1,0) can be written in the form $f(z)dz$ for some meromorphic function $f(z)$; and evidently the principal parts of the Laurent expansion of that meromorphic function at each of its poles describes the residue class of that differential form in the quotient sheaf $\mathcal{P}^{1,0}$. These local expansions can thus be used to describe elements of the sheaf $\mathcal{P}^{1,0}$. Recall that the residue of a meromorphic differential form φ at one of its poles p is defined to be the integral of φ around a closed path encircling that pole once in a counterclockwise direction and containing no other singularities of φ ; this residue, which will be denoted $2\pi i \, \mathcal{R}_p[\varphi]$, clearly remains unchanged by the addition to φ of any holomorphic differential form of type $(1,0)$, hence can be extended to be defined on the elements of the sheaf $\mathcal{P}^{1,0}$.

From the exact sequence of sheaves

$$0 \longrightarrow \mathcal{O}^{1,0} \longrightarrow \mathcal{M}^{1,0} \longrightarrow \mathcal{P}^{1,0} \longrightarrow 0$$

there results an exact cohomology sequence including the segment

$$\Gamma(M, \mathcal{M}^{1,0}) \longrightarrow \Gamma(M, \mathcal{P}^{1,0}) \xrightarrow{\delta} H^1(M, \mathcal{O}^{1,0}) \ ;$$

hence given any collection of principal parts $\varphi \in \Gamma(M, \mathcal{P}^{1,0})$ on the surface M , there exists a meromorphic Abelian differential on M having those principal parts as its singularities precisely when $\delta\varphi = 0 \in H^1(M, \mathcal{O}^{1,0})$. From the Serre duality theorem it follows that $H^1(M, \mathcal{O}^{1,0})$ is canonically dual to $H^0(M, \mathcal{O})$, and $H^0(M, \mathcal{O}) \cong \mathbf{C}$ for a compact Riemann surface; thus $H^1(M, \mathcal{O}^{1,0})$ can be identified with the complex numbers $\mathbf{C}$. Indeed this identification can be so chosen that the element $\delta\varphi \in H^1(M, \mathcal{O}^{1,0})$ is equated to the total residue of $\varphi \in \Gamma(M, \mathcal{P}^{1,0})$, leading thereby to the following result.

Theorem 2. On a compact Riemann surface M there exist meromorphic Abelian differentials with a prescribed principal part $\varphi \in \Gamma(M, \mathcal{P}^{1,0})$ if and only if the total residue of φ is zero.

Proof. As noted, it is only necessary to show that $\delta\varphi \in H^1(M, \mathcal{O}^{1,0})$ can be identified with the total residue of φ under an appropriate isomorphism $H^1(M, \mathcal{O}^{1,0}) \cong \mathbf{C}$. Choose a covering $\mathcal{U} = \{U_\alpha\}$ of M by contractible open coordinate neighborhoods, such that each singularity of φ is contained within only one of the sets U_α, indeed such that an open neighborhood of any singularity of φ contained within U_α is disjoint from the other sets of the covering. In each of these sets U_α the section $\varphi \in \Gamma(M, \mathcal{P}^{1,0})$ can be represented by a meromorphic function $\varphi_\alpha \in \Gamma(M, \mathcal{M}^{1,0})$; and in each intersection $U_\alpha \cap U_\beta$ the differences $\varphi_{\alpha\beta} = \varphi_\beta - \varphi_\alpha$ are holomorphic differential forms, and the cocycle $\{\varphi_{\alpha\beta}\} \in Z^1(\mathcal{U}, \mathcal{O}^{1,0})$ represents the cohomology class $\delta\varphi \in H^1(M, \mathcal{O}^{1,0})$ Then to apply Serre duality it is enough to choose C^∞ differential forms $\psi_\alpha \in \Gamma(U_\alpha, \mathcal{E}^{1,0})$ such that $\varphi_{\alpha\beta} = \psi_\beta - \psi_\alpha$ in $U_\alpha \cap U_\beta$; for $\bar\partial\psi_\alpha = \bar\partial\psi_\beta$ in $U_\alpha \cap U_\beta$, so these differential forms represent a global form $\{\bar\partial\psi_\alpha\} \in \Gamma(M, \mathcal{E}^{1,1})$, and the isomorphism $H^1(M, \mathcal{O}^{1,0}) \cong \mathbf{C}$ can be taken as that associating to the cocycle $\{\varphi_{\alpha\beta}\}$ the complex constant $\int_M \{\bar\partial\psi_\alpha\}$. Now for each set U_α choose a C^∞ function ε_α which is identically zero in an open neighborhood of any singularity of φ in U_α and which is identically one in $U_\alpha \cap U_\beta$ whenever $\alpha \neq \beta$. It is then possible to take $\psi_\alpha = \varepsilon_\alpha \varphi_\alpha$; and

$$\int_{U_\alpha} \bar\partial\psi_\alpha = \int_{U_\alpha} d\psi_\alpha = \int_{\partial U_\alpha} \psi_\alpha = \int_{\partial U_\alpha} \varphi_\alpha = 2\pi i \, \mathcal{R}_{U_\alpha}[\varphi_\alpha] ,$$

so that $\delta\varphi$ is identified with the total residue of φ over M and the proof is thereby concluded.

For any principal part $\varphi \in \Gamma(M, \mathcal{P}^{1,0})$ having total residue zero there exist meromorphic Abelian differentials having the singularities specified by φ. It is customary to call these forms Abelian differentials of the <u>second kind</u> if their residue at each point of M is zero, and Abelian differentials of the <u>third kind</u> if their total residue is zero but they have nonzero residues at some points. The holomorphic Abelian differentials are called the Abelian differentials of the <u>first kind</u>.

It is in some ways convenient to begin the more detailed discussion of meromorphic Abelian differentials by considering Abelian differentials of the third kind; and the simplest such differentials are those having two simple poles with opposite residues. Selecting any two distinct points p_-, p_+ on M, there exists a meromorphic Abelian differential $\varphi \in \Gamma(M, \mathcal{M}^{1,0})$ having as sole singularities simple poles at p_- with residue -1 and at p_+ with residue $+1$. The most general such meromorphic Abelian differential is of course $\varphi + \omega$ for an arbitrary holomorphic Abelian differential $\omega \in \Gamma(M, \mathcal{O}^{1,0})$ and a unique canonical such form can be specified by imposing some suitable conditions on the periods. Some care must be taken in defining the periods of such a differential form, though, since it has nonzero residues; but the difficulties can easily be avoided as follows. Select a simple arc δ from p_- to p_+ on M, and let $\tilde{\delta}$ be any lifting of this arc to the universal covering space $\tilde{M}$. The differential form φ can be viewed as a Γ-invariant meromorphic differential form on $\tilde{M}$; and as such it is a Γ-invariant holomorphic differential form on $\tilde{M} - \Gamma\tilde{\delta}$, and has zero integral around any closed path in $\tilde{M} - \Gamma\tilde{\delta}$. This form can therefore be written as the exterior derivative of a holomorphic function f on $\tilde{M} - \Gamma\tilde{\delta}$, the function f being unique up to an additive constant. Now as in the case of the holomorphic Abelian differentials it follows that $f(Tz) = f(z) - \varphi_\delta(T)$ for some complex constants $\varphi_\delta(T)$ depending on $T \in \Gamma$; and the set of these constants can be viewed as an element $\varphi_\delta \in \mathrm{Hom}(\Gamma, \mathbf{C})$, which will be called the <u>period class</u> of the meromorphic Abelian differential φ with respect to the arc δ. Having chosen a marking of the Riemann surface M , this period class is of course determined uniquely by its values $\varphi(A_i)$, $\varphi(B_i)$ on the canonical generators of Γ . There is a unique holomorphic Abelian differential also having the periods $\varphi(A_i)$ on the canonical generators $A_1, \ldots, A_g$; hence there is a unique differential form $\varphi + \omega$ having zero periods on the canonical generators $A_1, \ldots, A_g$. This will be called the <u>canonical Abelian differential of the third kind associated to the arc</u> δ , and will be denoted ω_δ . Thus $\omega_\delta \in \Gamma(M, \mathcal{M}^{1,0})$ is the unique meromorphic Abelian differential on M having as sole singularities simple poles at the points p_- with residue -1 and p_+ with residue $+1$, where δ is an arc from p_- to p_+ , and having period class $\omega_\delta \in \mathrm{Hom}(\Gamma, \mathbf{C})$ such that $\omega_\delta(A_i) = 0$ for $1 \leqq i \leqq g$. This condition on the period class can be expressed equivalently as the condition that $\int_\gamma \omega_\delta = 0$ along any one-cycle γ homologous to some linear combination of the one-cycles $\alpha_1,\ldots,\alpha_g$ and disjoint from the arc δ. Correspondingly the canonical Abelian integral of the third kind associated to the arc δ is the unique holomorphic function w_δ on $\tilde{M} - \Gamma\tilde{\delta}$ such that $dw_\delta = \omega_\delta$ and $w_\delta(z_o) = 0$. Note though that this is not really defined when $p_o \in \delta$, but only because the normalization condition $w_\delta(z_o) = 0$ is not well defined; there certainly exist holomorphic functions on $\tilde{M} - \Gamma\tilde{\delta}$ having exterior derivative equal to ω_δ in this case as well. Note that the function $w_\delta(z)$ can evidently be continued analytically across any interior point of one of the arcs $\Gamma\tilde{\delta}$, but has logarithmic branch points at the points of $\tilde{M}$ lying over p_+ and p_-, hence the continuation is multiple valued on $\tilde{M}$. The Abelian integral $w_\delta(z)$ satisfies the equation $w_\delta(Tz) = w_\delta(z) - \omega_\delta(T)$, where $\omega_\delta \in \mathrm{Hom}(\Gamma,\mathbf{C})$ is the period class of that canonical Abelian differential. The natural extension of Riemann's equality leads to the following result.

Theorem 3. (a) For a canonical Abelian differential of the third kind ω_δ on a marked Riemann surface, so long as the arc δ is disjoint from the cuts associated to the marking,

$$\omega_\delta(B_j) = 2\pi i \int_\delta \omega_j \qquad \text{for } j = 1,\ldots,g .$$

(b) For any pair of canonical Abelian differentials of the third kind $\omega_{\delta'}, \omega_{\delta''}$ on a marked Riemann surface, so long as the arcs δ' and δ'' are disjoint from one another and from the cuts associated to the marking,

$$\int_{\delta'} \omega_{\delta''} = \int_{\delta''} \omega_{\delta'} .$$

Proof. (a) Let $\tilde{\delta} \subset \tilde{M}$ be that lifting of the arc δ which is contained within the canonical fundamental domain Δ, and let z_-, z_+ be the initial and terminal points of $\tilde{\delta}$ respectively. Then viewed as a Γ-invariant meromorphic differential form on $\tilde{M}$ the only singularities of ω_δ in $\bar{\Delta}$ are simple poles at z_+ with residue $+1$ and at z_- with residue -1. It therefore follows that for any holomorphic Abelian differential $\omega_j = dw_j$,

$$\int_{\partial\Delta} w_j \omega_\delta = 2\pi i \; \mathcal{R}_\Delta[w_j \omega_\delta] = 2\pi i[w_j(z_+) - w_j(z_-)]$$

$$= 2\pi i \int_{\tilde{\delta}} \omega_j = 2\pi i \int_\delta \omega_j ,$$

where as usual $\mathcal{R}_\Delta[\;]$ denotes the total residue in Δ. On the other hand the boundary integral can be calculated in terms of the period classes of the two differential forms, exactly as in the proof of Riemann's equality in Theorem 1; referring to that proof for the details of the calculation, and recalling that the periods of these differential forms are normalized so that $\omega_j(A_k) = \delta^j_k$ and $\omega_\delta(A_k) = 0$, it follows that

$$\int_{\partial\Delta} w_j \omega_\delta = \sum_{k=1}^{g} \left\{ \omega_j(A_k)\omega_\delta(B_k) - \omega_j(B_k)\omega_\delta(A_k) \right\} = \omega_\delta(B_j) ,$$

proving the first assertion of the theorem.

(b) Let $\tilde{\delta}'$ and $\tilde{\delta}''$ be those liftings of the arcs δ' and δ'' respectively which are contained in Δ, and let z'_-, z''_- be the initial points and z'_+, z''_+ be the terminal points of these liftings; and let σ' and σ'' be disjoint closed paths in Δ encircling $\tilde{\delta}'$ and $\tilde{\delta}''$ respectively. Since the differential form $w_{\delta'}\omega_{\delta''}$ is holomorphic in $\Delta - (\tilde{\delta}' \cup \tilde{\delta}'')$ it follows that

$$\int_{\partial\Delta} w_{\delta'}\omega_{\delta''} = \int_{\sigma'} w_{\delta'}\omega_{\delta''} + \int_{\sigma''} w_{\delta'}\omega_{\delta''} .$$

Now the function $w_{\delta'}$ is holomorphic in the interior of σ'', while the differential form $\omega_{\delta''}$ is meromorphic with simple poles at z''_+ with residue $+1$ and at z''_- with residue -1 ; therefore

$$\int_{\sigma''} w_{\delta'}\omega_{\delta''} = 2\pi i[w_{\delta'}(z''_+) - w_{\delta'}(z''_-)] = 2\pi i \int_{\tilde{\delta}''} \omega_{\delta'} = 2\pi i \int_{\delta''} \omega_{\delta'} .$$

On the other hand both functions $w_{\delta'}$, $w_{\delta''}$ are C^∞ along the closed path σ', so that by Stokes' theorem

$$\int_{\sigma'} w_{\delta'}\omega_{\delta''} = \int_{\sigma'} [d(w_{\delta'}w_{\delta''}) - w_{\delta''}\omega_{\delta'}] = - \int_{\sigma'} w_{\delta''}\omega_{\delta'} ;$$

and it then follows as above that

$$\int_{\sigma'} w_{\delta''}\omega_{\delta'} = 2\pi i \int_{\delta'} \omega_{\delta''} .$$

Thus altogether

$$\int_{\partial\Delta} w_{\delta'}\omega_{\delta''} = 2\pi i \int_{\delta''} \omega_{\delta'} - 2\pi i \int_{\delta'} \omega_{\delta''} .$$

Again using the calculation from the proof of Theorem 1, and recalling the normalizations $\omega_{\delta'}(A_k) = \omega_{\delta''}(A_k) = 0$, it follows that

$$\int_{\partial\Delta} w_{\delta'}\omega_{\delta''} = \sum_{k=1}^{g} \left\{ \omega_{\delta'}(A_k)\omega_{\delta''}(B_k) - \omega_{\delta'}(B_k)\omega_{\delta''}(A_k) \right\} = 0 ,$$

concluding the proof of the theorem.

(d) Let δ be a simple arc from p_- to p_+ on a marked Riemann surface M, such that δ is disjoint from the canonical cuts associated to the marking; and let $\tilde{\delta}$ be the simple arc from z_- to z_+ in the canonical fundamental domain Δ, lying over δ. The canonical Abelian integral of the third kind w_δ is holomorphic on $\tilde{M} - \Gamma\tilde{\delta}$, and can be continued analytically across each arc $\Gamma\tilde{\delta}$ but with logarithmic branching at the end points of the arcs; since $\omega_\delta = dw_\delta$ has residues ± 1 at these end points, it follows that the exponential $p_\delta = e^{w_\delta}$ extends to a single-valued meromorphic function on all of $\tilde{M}$, with simple zeros at the points Γz_+ and simple poles at the points Γz_-, but otherwise holomorphic and nowhere vanishing. Note that $p_\delta(Tz) = \chi_\delta(T)p_\delta(z)$ for every $T \in \Gamma$, where $\chi_\delta \in \mathrm{Hom}(\Gamma, \mathbf{C}^*)$; this representation χ_δ is determined by its values on the canonical generators of Γ, and it follows from the definition of the canonical Abelian differentials of the third kind that $\chi_\delta(A_k) = 1$ for $k = 1,\ldots,g$, and from Theorem 3(a) that $\chi_\delta(B_k) = \exp 2\pi i \int_\delta \omega_k$ for $k = 1,\ldots,g$.

As an incidental observation, suppose given any set of $2r$ points $p_1,\ldots,p_r,q_1,\ldots,q_r$ on the surface M, and select simple arcs δ_j from q_j to p_j avoiding some set of canonical cuts associated to a marking of the surface. The function

$$p(z) = \prod_{j=1}^{r} p_{\delta_j}(z)$$

is then meromorphic on all of $\tilde{M}$, has simple zeros at the points p_i and simple poles at the points q_i (or multiple zeros and poles appropriately if these points are not distinct) but is otherwise holomorphic and nonvanishing, and satisfies $p(Tz) = \chi(T)p(z)$ for a representation $\chi \in \mathrm{Hom}(\Gamma, \mathbf{C}^*)$ determined by

$$\chi(A_k) = 1 , \quad \chi(B_k) = \exp 2\pi i \sum_{j=1}^{r} \int_{\delta_j} \omega_k \quad (\text{for } k = 1,\ldots,g).$$

This shows quite constructively that the complex line bundle $\zeta_{p_1}\cdots\zeta_{p_r}\zeta_{q_1}^{-1}\cdots\zeta_{q_r}^{-1}$ (where ζ_p denotes a point bundle) is analytically equivalent to the flat line bundle associated to the representation χ , hence provides an alternative proof of Abel's theorem (Theorem 18 of the earlier lecture notes). In particular, it follows readily that there exists a meromorphic function on the surface M with divisor $p_1 + \ldots + p_r - q_1 - \ldots - q_r$ precisely when there exists a holomorphic Abelian differential ω such that

$$\chi(A_k) = \exp \omega(A_k) , \quad \chi(B_k) = \exp \omega(B_k) , \quad (\text{for } k = 1,\ldots,g);$$

and that function is $p(z)e^{-w(z)}$ where $dw(z) = \omega(z)$. This reduces to the more traditional form of Abel's theorem.

Next let σ be another simple arc on the marked Riemann surface M , say from q_- to q_+ , such that σ is disjoint from δ and from the canonical cuts associated to the marking; and let $\tilde{\sigma}$ be the simple arc from t_- to t_+ in the canonical fundamental domain Δ , lying over σ . It then follows from Theorem 3(b) that $\int_\delta \omega_\sigma = \int_\sigma \omega_\delta$, or in terms of the canonical Abelian integrals that

$w_\sigma(z_+) - w_\sigma(z_-) = w_\delta(t_+) - w_\delta(t_-)$; and consequently

$$\frac{p_\sigma(z_+)}{p_\sigma(z_-)} = \frac{p_\delta(t_+)}{p_\delta(t_-)} .$$

Now for any four distinct points z_+, z_-, t_+, t_- in Δ introduce the function defined by

$$(8) \qquad p(z_+,z_-,t_+,t_-) = \frac{p_\sigma(z_+)}{p_\sigma(z_-)} = \frac{p_\delta(t_+)}{p_\delta(t_-)} ,$$

noting that it depends only on these four points and not on the choices of the arcs σ, δ . It is evident from this definition that, holding any three of these points fixed, the function $p(z_+, z_-, t_+, t_-)$ extends to a meromorphic function on all of $\tilde{M}$; hence $p(z_+, z_-, t_+, t_-)$ is a meromorphic function of the four complex variables (z_+, z_-, t_+, t_-) on the complex analytic manifold $\tilde{M}^4 = \tilde{M} \times \tilde{M} \times \tilde{M} \times \tilde{M}$. This will be called the <u>prime function</u> on the marked Riemann surface M ; the characteristic properties of this prime function, which follow immediately from the definition, are summarized as follows.

<u>Theorem 4.</u> The prime function on a marked Riemann surface M is the meromorphic function $p(z_+, z_-, t_+, t_-)$ on $\tilde{M}^4$ characterized by the following properties:

(i) $p(z_+, z_-, t_+, t_-) = p(t_+, t_-, z_+, z_-) = p(z_-, z_+, t_+, t_-)^{-1}$;

(ii) for fixed z_-, t_+, t_- the function $p(z, z_-, t_+, t_-)$ has simple zeros at the points Γt_+ , simple poles at the points Γt_- , and is otherwise holomorphic and nonzero on $\tilde{M}$, provided that $\Gamma t_+ \neq \Gamma t_-$; and this function is identically one if $t_+ = t_-$;

(iii) for any transformation $T \in \Gamma$

$$p(Tz_+, z_-, t_+, t_-) = \chi(T; t_+, t_-)\cdot p(z_+, z_-, t_+, t_-) ,$$

where $\chi(\cdot\ ; t_+, t_-) \in \mathrm{Hom}(\Gamma, \mathbf{C}^*)$ *is the representation defined by*

$$\chi(A_k; t_+, t_-) = 1 , \quad \chi(B_k; t_+, t_-) = \exp 2\pi i[w_k(t_-) - w_k(t_+)]$$

for $k = 1,\ldots,g$.

Proof. That the prime function satisfies these conditions follows almost immediately from the definition. Conversely it follows equally trivially that any function satisfying these three conditions must be a constant multiple of the prime function, since the quotient is evidently holomorphic, nowhere vanishing, and Γ-invariant on $\tilde{M}^4$; indeed that constant must be one, as a consequence of the last part of condition (ii).

Note that the canonical Abelian differential of the third kind can readily be recovered from the prime function on the marked surface M, since

$$(9) \qquad \omega_\delta(z) = d \log p(z, z_-, t_+, t_-)$$

where t_-, t_+ are the initial and terminal points of any lifting of the arc δ to the covering space $\tilde{M}$ and z_- is any point of $\tilde{M}$ not contained in $\Gamma t_+ \cup \Gamma t_-$; indeed the canonical Abelian integral of the third kind is evidently given by

(10) $$w_\delta(z) = \log p(z, z_o, t_+, t_-) ,$$

choosing that branch of the logarithm in $\tilde{M} - \Gamma\tilde{\delta}$ which vanishes at the base point z_o .

Other canonical meromorphic Abelian differentials can also be obtained from this prime function; the only complication is that these other differentials are not functions but differential forms or similar expressions in the parameters describing the location of their singularities. Thus view $p(z, z_o, t, t_o)$ as a function of the two complex variables $(z,t) \in \tilde{M}^2$ for fixed points z_o, t_o ; and letting t also denote some local coordinate at the point $t \in \tilde{M}$, introduce the meromorphic function

(11) $$w_t(z) = \frac{\partial}{\partial t} \log p(z, z_o, t, t_o)$$

of the point $z \in \tilde{M}$. Note that this function is really just the coefficient of the canonical Abelian differential $\omega_\delta(t)$, where $\tilde{\delta}$ is an arc from z_o to z , when that differential form is written in terms of the local coordinate t as $\omega_\delta(t) = w_t(z)dt$. It is easily verified that $w_t(z_o) = 0$ and that $w_t(z)$ is completely independent of the choice of the point t_o ; that $w_t(z)$ has simple poles at the points Γt and is otherwise holomorphic on $\tilde{M}$; and that $w_t(Tz) = w_t(z) - \omega_t(T)$ where $\omega_t \in \mathrm{Hom}(\Gamma, \mathbf{C})$ is the homomorphism determined by

(12) $$\omega_t(A_k) = 0 , \quad \omega_t(B_k) = 2\pi i \frac{d}{dt} w_k(t) \quad \text{for } k = 1,\dots,g .$$

Indeed, letting z and t denote the complex coordinate of these points in the local coordinates chosen, it is further evident that the principal part of the function $w_t(z)$ is precisely $-\frac{1}{z-t}$. The differential form $\omega_t(z) = dw_t(z)$ is called the canonical Abelian differential of the second kind associated to the given coordinate system at the point t. This form has a double pole at the point t with zero residue; indeed in terms of the given coordinate system the principal part of this differential form is precisely $\frac{1}{(z-t)^2}\,dz$ at the point t. The period class of this form is the homomorphism $\omega_t \in \mathrm{Hom}(\Gamma,\mathbb{C})$, and the function $w_t(z)$ is the associated canonical Abelian integral of the second kind. Meromorphic differential forms with poles of higher order can be obtained similarly, merely taking higher derivatives of $\log p(z, z_o, t, t_o)$ with respect to the local coordinate t; and all of these of course then depend on the choice of local coordinate system.

Notes for §1.

(a) If one is willing to take as known the general uniformization theorem, the universal covering space $\tilde{M}$ of a compact Riemann surface M of genus $g > 0$ can be identified with a subset of the complex plane; indeed $\tilde{M}$ can be identified with the full complex plane if $g = 1$ or the unit disc if $g > 1$, and the covering transformation group Γ has a correspondingly explicit form. However this identification really plays no role whatsoever in most of the present discussion, and will generally be ignored here.

(b) The period classes of Abelian differentials can also be introduced as follows. From the exact sequence of sheaves

$$0 \longrightarrow \mathbf{C} \longrightarrow \mathcal{O} \xrightarrow{d} \mathcal{O}^{1,0} \longrightarrow 0$$

there follows an exact cohomology sequence, which for a compact Riemann surface reduces to the sequence

$$0 \longrightarrow \Gamma(M, \mathcal{O}^{1,0}) \xrightarrow{\delta} H^1(M,\mathbf{C}) \longrightarrow H^1(M, \mathcal{O}) \longrightarrow 0 ,$$

as noted in §8(a) of the earlier lecture notes. Now

$$H^1(M,\mathbf{C}) \cong \mathrm{Hom}(H_1(M,\mathbf{Z}),\mathbf{C}) \cong \mathrm{Hom}(\pi_1(M,p_0),\mathbf{C}) \cong \mathrm{Hom}(\Gamma,\mathbf{C}) ,$$

and the connecting homomorphism δ can thereby be identified with the mapping which assigns to any Abelian differential its period class. The remainder of this portion of the exact cohomology sequence appears as a condition determining which elements of

$H^1(M,\mathbf{C}) \cong \mathrm{Hom}(\Gamma,\mathbf{C})$ are the period classes of the Abelian differentials on M ; the condition is just that an element of $H^1(M,\mathbf{C})$ be mapped to zero under the homomorphism $H^1(M,\mathbf{C}) \longrightarrow H^1(M, \mathcal{O})$ induced by the inclusion $\mathbf{C} \longrightarrow \mathcal{O}$. The resulting condition is a triviality; but it is a good exercise to verify that that is the case by tracing this condition through in detail.

The algebraic condition (6) satisfied by the quadratic period classes should be familiar to anyone acquainted with the cohomology theory of abstract groups. That condition is just that the two-cocycle $c(S,T) = \omega_i(S)\omega_j(T) \in Z^1(\Gamma,\mathbf{C})$ is the coboundary of the one-cochain $\sigma_{ij}(T) \in C^1(\Gamma,\mathbf{C})$, in terms of the cohomology of the group Γ with coefficients in the trivial Γ-module $\mathbf{C}$. Actually the two-cocycle $c(S,T) = \omega_i(S)\omega_j(T)$ is just the cup product of the one-cocycles $\omega_i(T)$ and $\omega_j(T)$, and the condition that this two-cocycle be cohomologous to zero is the algebraic reflection of the fact that the differential form $\omega_i(z)\wedge\omega_j(z)$ is cohomologous to zero on M ; so the algebraic condition that there exist some mappings $\sigma_{ij}(T)$ satisfying (6) is really equivalent to Riemann's equality.

(c) That a function of several complex variables is meromorphic if it is meromorphic in each variable separately is a result of W. Rothstein, extending the well known theorem of Osgood and Hartogs with the corresponding assertion for holomorphic functions. For a survey of these results, see H. Behnke & P. Thullen, Theorie der Funktionen mehrerer komplexer Veränderlichen (second edition, Springer, 1970).

If $\tilde{M}$ is identified with a subset of the complex plane, then points of $\tilde{M}$ can be viewed as complex numbers and the canonical Abelian integral of the second kind can be defined simply as the derivative

$$(13) \qquad w_t(z) = \frac{\partial}{\partial t} \log p(z,z_o,t,t_o)$$

and the corresponding canonical Abelian differential as

$$(14) \qquad \omega_t(z) = \frac{\partial^2}{\partial z \partial t} \log p(z,z_o,t,t_o)dz \ .$$

This is in many ways a great convenience, especially when considering Abelian differentials of the second kind with higher order poles; and the principal parts of meromorphic Abelian differentials can then be described without the need for specifying a choice of local coordinate system. The canonical Abelian differentials of the second kind can also be viewed as a symmetric differential form

$$\Omega(z,t) = \frac{\partial^2}{\partial z \partial t} \log p(z,z_o,t,t_o)dz \otimes dt$$

on the product manifold $M \times M$, and the period class correspondingly as the homomorphism $\Omega \in \mathrm{Hom}(\Gamma,\Gamma(M, \mathcal{O}^{1,0}))$ described by

$$\Omega(A_k) = 0 \ , \qquad \Omega(B_k) = 2\pi i \ \omega_k \qquad \text{for} \quad k = 1,\dots,g \ .$$

For such a simple situation as this, though, it seems unnecessary to worry unduly with such additional machinery.

When considering meromorphic functions rather than meromorphic differential forms, the appropriate sheaf of germs of principal parts is the quotient sheaf $\mathcal{P} = \mathcal{M}/\mathcal{O}$; and from the exact sequence of sheaves

$$0 \longrightarrow \mathcal{O} \longrightarrow \mathcal{M} \longrightarrow \mathcal{P} \longrightarrow 0$$

there results an exact cohomology sequence including the segment

$$\Gamma(M, \mathcal{M}) \longrightarrow \Gamma(M, \mathcal{P}) \xrightarrow{\delta} H^1(M, \mathcal{O}) .$$

In this case the Serre duality theorem asserts that $H^1(M, \mathcal{O})$ is canonically dual to $\Gamma(M, \mathcal{O}^{1,0})$; and it is easy to verify, paralleling the proof of Theorem 2, that this pairing is that which associates to the section $f \in \Gamma(M, \mathcal{P})$ and to any holomorphic Abelian differential $\omega \in \Gamma(M, \mathcal{O}^{1,0})$ the complex number

$$(\delta f, \omega) = 2\pi i \ \mathcal{R}_M[f \cdot \omega] .$$

Consequently there exists a global meromorphic function on M with principal part $f \in \Gamma(M, \mathcal{P})$ if and only if $\mathcal{R}_M[f\ \omega_j] = 0$ for $j = 1, \ldots, g$, in terms of any basis for the space of Abelian differentials $\Gamma(M, \mathcal{O}^{1,0})$. Actually the more detailed analysis of this result can be reduced directly to the study of the meromorphic differentials, though. For given any principal part $f \in \Gamma(M, \mathcal{P})$ note that $df \in \Gamma(M, \mathcal{P}^{1,0})$ necessarily has zero residue, hence there exists a global meromorphic differential form $\varphi \in \Gamma(M, \mathcal{M}^{1,0})$

having the principal part df ; and this form is unique up to an additive Abelian differential form $\omega \in \Gamma(M, \mathcal{O}^{1,0})$. Then there exists a global meromorphic function with the principal part f if and only if there exists an Abelian differential $\omega \in \Gamma(M, \mathcal{O}^{1,0})$ such that the form $\varphi + \omega$ has zero period class. The verification that this condition is equivalent to the condition obtained above from the Serre duality theorem is an interesting exercise that can be left to the reader, with the proof of Theorem 3 as a hint if one is necessary.

§2. Jacobi varieties and their distinguished subvarieties.

(a) Consider once again the canonical period matrix (I,Ω) of a marked Riemann surface M, where $I = \{\delta^i_j\}$ is the $g \times g$ identity matrix and $\Omega = \{\omega_{ij}\}$ is the $g \times g$ matrix whose entries $\omega_{ij} = \omega_i(B_j)$ are the periods of the canonical holomorphic Abelian differentials. The columns of this matrix can be viewed as a set of $2g$ vectors in $\mathbb{C}^g$, and they are linearly independent over the real numbers as a consequence of Riemann's inequality; hence these vectors generate a lattice subgroup $\mathcal{L} \subset \mathbb{C}^g$, such that the quotient space $\mathbb{C}^g/\mathcal{L}$ is a compact complex torus. (For more details, see for instance the discussion in §8(b) of the earlier lecture notes.) This manifold will be called the <u>Jacobi variety</u> of the marked Riemann surface M, and will be denoted by $J(M)$. Note that the vector $u = \{u_i\} \in \mathbb{C}^g$ belongs to the lattice subgroup $\mathcal{L}$ precisely when there are integers $m_1,\ldots,m_g,n_1,\ldots,n_g$ such that

$$u_i = \sum_{j=1}^{g} [\delta^i_j m_j + \omega_{ij} n_j] = \sum_{j=1}^{g} [m_j\omega_i(A_j) + n_j\omega_i(B_j)]$$

$$= \omega_i(A_1^{m_1} \cdots A_g^{m_g} B_1^{n_1} \cdots B_g^{n_g}) \qquad \text{for } i = 1,\ldots,g ,$$

where $\omega_i \in \mathrm{Hom}(\Gamma,\mathbb{C})$ is the period class of the holomorphic Abelian differential $\omega_i \in \Gamma(M, \mathcal{O}^{1,0})$; thus the lattice $\mathcal{L}$ consists of all vectors $u = \{\omega_i(T)\} \in \mathbb{C}^g$ for elements $T \in \Gamma$.

The values $\{w_i(z)\}$ of the canonical holomorphic Abelian integrals at a point $z \in \tilde{M}$ can be viewed as the components of a vector $\tilde{\varphi}(z) \in \mathbb{C}^g$, thus defining a complex analytic mapping $\tilde{\varphi}: \tilde{M} \longrightarrow \mathbb{C}^g$. Note that for any covering transformation $T \in \Gamma$,

$$\tilde{\varphi}(Tz) - \tilde{\varphi}(z) = \{w_i(Tz) - w_i(z)\} = \{\omega_i(T^{-1})\} \in \mathcal{L} ;$$

consequently $\tilde{\varphi}$ induces a complex analytic mapping $\varphi: M \longrightarrow J(M)$, which will be called the Jacobi mapping of the marked Riemann surface M into its Jacobi variety.

Theorem 5. For a Riemann surface M of positive genus the Jacobi mapping $\varphi: M \longrightarrow J(M)$ is a complex analytic homeomorphism between M and a complex analytic submanifold $W_1 \subseteq J(M)$.

Proof. The Jacobian of the mapping φ is the vector consisting of the canonical holomorphic Abelian differentials on M , and is nonsingular since these differentials have no common zeros at any point of M (as observed on page 119 of the earlier lecture notes). Since M is compact, it follows immediately that the image $W_1 = \varphi(M) \subseteq J(M)$ is a complex analytic submanifold of $J(M)$, and that the mapping $\varphi: M \longrightarrow W_1$ is locally a complex analytic homeomorphism. It is a simple consequence of Abel's theorem that the mapping φ is one-to-one. For if $\varphi(p_1) = \varphi(p_2)$ for two points p_1, p_2 on M and δ is any simple arc from p_1 to p_2 , then the vector $u = \{u_i\}$ with components $u_i = \int_\delta \omega_i$ belongs to the lattice $\mathcal{L}$, and consequently $\zeta_{p_1} = \zeta_{p_2}$ (upon recalling Corollary 2 to Theorem 18 of the earlier lecture notes); but $\zeta_{p_1} = \zeta_{p_2}$ for a surface of positive genus implies that $p_1 = p_2$ (upon recalling Lemma 16 of the earlier lecture notes and the subsequent discussion).

This then serves to conclude the proof.

Since the Jacobi variety $J(M)$ has a natural group structure, it is evident that the Jacobi mapping $\varphi: M \longrightarrow J(M)$ extends uniquely to a homomorphism from the free Abelian group generated by the points of M to $J(M)$. Of course this free Abelian group on the points of M can be identified with the <u>group of divisors</u> on M, the group $\Gamma(M, \mathcal{D})$ where $\mathcal{D} = \mathcal{M}^*/\mathcal{O}^*$ is the sheaf of germs of divisors on M; and the group homomorphism $\varphi: \Gamma(M, \mathcal{D}) \longrightarrow J(M)$ is clearly given by $\varphi(\Sigma_i \nu_i p_i) = \Sigma_i \nu_i \varphi(p_i)$, where $\nu_i \in Z$, $p_i \in M$. Recall that two divisors $\mathfrak{d}_1$, $\mathfrak{d}_2$ in $\Gamma(M, \mathcal{D})$ are called <u>linearly equivalent,</u> written $\mathfrak{d}_1 \approx \mathfrak{d}_2$, if their difference is the divisor of a meromorphic function on M; that this is the equivalence relation naturally corresponding to the homomorphism $\Gamma(M, \mathcal{D}) \longrightarrow H^1(M, \mathcal{O}^*)$ which associates to a divisor $\mathfrak{d} = \Sigma_i \nu_i p_i \in \Gamma(M, \mathcal{D})$ the complex line bundle $\zeta_{\mathfrak{d}} = \Pi_i \zeta_{p_i}^{\nu_i} \in H^1(M, \mathcal{O}^*)$, where ζ_{p_i} is the point bundle associated to $p_i \in M$; and that the group of linear equivalence classes of divisors on M, the <u>divisor class group</u> of M, is isomorphic to the group $H^1(M, \mathcal{O}^*)$ of complex line bundles over M under the induced homomorphism. Note that the Chern class $c(\zeta_{\mathfrak{d}})$ of the line bundle $\zeta_{\mathfrak{d}}$ associated to the divisor $\mathfrak{d} = \Sigma_i \nu_i p_i$ is just equal to the <u>degree</u> $|\mathfrak{d}| = \Sigma_i \nu_i$ of that divisor.

Theorem 6. For a Riemann surface M of positive genus the Jacobi homomorphism

$$\varphi\colon \Gamma(M, \mathcal{D}) \longrightarrow J(M)$$

is surjective, and its kernel consists of those divisors $\mathfrak{d} \in \Gamma(M, \mathcal{D})$ such that $\mathfrak{d} \approx |\mathfrak{d}| p_o$, where $p_o \in M$ is the base point of M.

Proof. Note that for the base point $p_o \in M$ it follows from the definition of the Jacobi mapping that $\varphi(p_o) = 0 \in J(M)$; and since φ is a group homomorphism, $\varphi(\mathfrak{d} - |\mathfrak{d}|\cdot p_o) = \varphi(\mathfrak{d}) - |\mathfrak{d}|\cdot\varphi(p_o) = \varphi(\mathfrak{d})$ for any divisor $\mathfrak{d}$. Now the divisor $\mathfrak{d} - |\mathfrak{d}| p_o$ has degree zero, so can be written in the form $\mathfrak{d} - |\mathfrak{d}|\cdot p_o = \Sigma_j (p_j^+ - p_j^-)$ for some points p_j^+, p_j^- of M; and selecting arcs δ_j from p_j^- to p_j^+ in M, it is evident that $\varphi(\mathfrak{d} - |\mathfrak{d}|\cdot p_o) \in J(M) = \mathbf{C}^g/\mathcal{L}$ is represented by the vector $u \in \mathbf{C}^g$ having components $u_k = \Sigma_j \int_{\delta_j} \omega_k$ for $k = 1,\ldots,g$. Then $\varphi(\mathfrak{d}) = 0$ precisely when $u \in \mathcal{L}$, and by Abel's theorem that is in turn equivalent to the condition that $\mathfrak{d} - |\mathfrak{d}|\cdot p_o \approx 0$; hence the kernel of φ has the form asserted. For a divisor $\mathfrak{d}$ of degree $|\mathfrak{d}| = 0$ it follows in particular that $\varphi(\mathfrak{d}) = 0$ precisely when $\mathfrak{d} \approx 0$, or equivalently precisely when the associated line bundle $\zeta_{\mathfrak{d}}$ is analytically trivial; consequently the restriction of φ to the subgroup of divisors of degree zero induces an injective homomorphism from the Picard variety $P(M) = \{\xi \in H^1(M, \mathcal{O}^*) | c(\xi) = 0\}$ to the Jacobi variety $J(M)$.

Since both $P(M)$ and $J(M)$ are complex tori of dimension g this homomorphism must evidently be an isomorphism; so of course φ is surjective, and the proof is thereby completed.

As an immediate corollary of this result, note that the mapping which associates to any divisor $\mathfrak{d} \in \Gamma(M, \mathcal{D})$ the pair $\varphi(\mathfrak{d}) \in J(M)$, $|\mathfrak{d}| \in \mathbf{Z}$, is a surjective homomorphism from $\Gamma(M, \mathcal{D})$ to the direct sum group $J(M) \oplus \mathbf{Z}$, and its kernel consists precisely of those divisors linearly equivalent to zero; this homomorphism thus determines an isomorphism between the divisor class group of M and the group $J(M) \oplus \mathbf{Z}$. Recalling that the homomorphism $\Gamma(M, \mathcal{D}) \longrightarrow H^1(M, \mathcal{O}^*)$ associating to a divisor $\mathfrak{d} = \Sigma_j \nu_j p_j$ the line bundle $\zeta_{\mathfrak{d}} = \Pi_j \zeta_{p_j}^{\nu_j}$ induces a canonical isomorphism between the divisor class group and the group of line bundles over M, the composition of these two isomorphisms leads to a canonical isomorphism

$$H^1(M, \mathcal{O}^*) \cong J(M) \oplus \mathbf{Z} .$$

In particular, there results a canonical isomorphism between the Picard variety $P(M) \cong \{\xi \in H^1(M, \mathcal{O}^*) | c(\xi) = 0\}$ and the Jacobi variety $J(M)$; and this canonical isomorphism will be used quite freely in the sequel to identify the Picard variety $P(M)$ with the Jacobi variety $J(M)$. This is of course really just an explicit form for the isomorphism discussed in §8 of the earlier notes; it can be viewed as providing an explicit coordinate representation of the Picard variety as a complex torus. At the same time it demonstrates that the Jacobi variety as an analytic group is really independent of the marking on the surface, except for the choice of base point on the surface and corresponding identity element of the group.

Note that the image under the Jacobi mapping $\varphi: M \longrightarrow J(M)$ of a point $p \in M$ is the same as the image under the Jacobi homomorphism $\varphi: \Gamma(M, \mathcal{D}) \longrightarrow J(M)$ of the divisor $p - p_o$; and this divisor corresponds to the line bundle $\zeta_p \zeta_{p_o}^{-1} \in P(M)$. Thus viewed as a subset of the Picard variety, W_1 consists of those line bundles $\xi \in P(M)$ which can be written in the form $\xi = \zeta_p \zeta_{p_o}^{-1}$ for some point $p \in M$. This is equivalent to the condition that the line bundle $\xi\zeta_{p_o}$ have a nontrivial holomorphic section; for $\zeta_p = \xi\zeta_{p_o}$ does have such a section, and conversely if $\xi\zeta_{p_o}$ has a nontrivial section with divisor p then $\xi\zeta_{p_o} = \zeta_p$. Writing $\gamma(\xi) = \dim_{\mathbb{C}} \Gamma(M, \mathcal{O}(\xi))$ for any complex line bundle ξ , it follows that

$$W_1 = \{\xi \in P(M) | \gamma(\xi\zeta_{p_o}) \geqq 1\} .$$

(b) The image under the Jacobi homomorphism $\varphi: \Gamma(M, \mathcal{D}) \longrightarrow J(M)$ of the set of positive divisors of degree r on the surface M is a well defined subset $W_r \subseteq J(M)$. This subset W_r can be described equivalently as the image of the complex analytic mapping $\varphi: M^r \longrightarrow J(M)$ defined by $\varphi(p_1, \ldots, p_r) = \varphi(p_1 + \ldots + p_r) = \varphi(p_1) + \ldots + \varphi(p_r)$, where M^r is the Cartesian product of r copies of the Riemann surface M and has the obvious structure of a compact complex analytic manifold of dimension r . It is a well known result in the theory of functions of several complex variables (the proper mapping theorem) that the image of any such mapping is a complex analytic subvariety of the complex manifold $J(M)$; hence the subsets $W_r \subseteq J(M)$ are complex analytic subvarieties. Actually these are also irreducible subvarieties, in the sense that any meromorphic function on $J(M)$ which vanishes on a relatively open subset of W_r necessarily vanishes identically on W_r ; that is an immediate consequence of the identity theorem for functions of several complex variables, since the restriction to W_r of a meromorphic function on $J(M)$ can be identified with a meromorphic function on the manifold M^r by means of the mapping φ . In terms of local coordinates z_j near the points $p_j \in M$, the Jacobian of the analytic mapping $\varphi: M^r \longrightarrow J(M)$ at the point $(z_1,\dots,z_r) \in M^r$ is clearly the $r \times g$ matrix $\{w_i'(z_j)\}$, where $\omega_i(z_j) = w_i'(z_j)dz_j$ are the canonical holomorphic Abelian differentials on M expressed in these local coordinates; and since the Abelian differentials are linearly independent, it is evident that the Jacobian matrix will have maximal rank on a dense open subset of M^r , indeed on the complement of a proper complex analytic subvariety of M^r . (See the discussion on page 119 of the preceding lecture notes.) In particular, for $r = 1,\dots,g$, the mapping $\varphi: M^r \longrightarrow J(M)$ will be a nonsingular local homeomorphism on a dense open subset of the manifold M^r ; and calling once again on some results from the theory of functions of several complex variables, it follows that the irreducible subvarieties $W_r \subseteq J(M)$ are of dimension r, for $r = 1,\ldots,g$. Of course for $r = 1$ it has already been demonstrated that the Jacobi mapping $\varphi: M \longrightarrow J(M)$ is everywhere nonsingular, is indeed a complex analytic homeomorphism between M and W_1, so that W_1 is an analytic submanifold of $J(M)$. For $r = g$ it follows that $W^g = J(M)$, since $J(M)$ is a manifold (hence irreducible) and dimension W_g = dimension $J(M)$; this assertion is traditionally known as the Jacobi inversion theorem. For $1 < r < g$ the mapping $\varphi: M^r \longrightarrow J(M)$ does have singularities, and the image W_r may also have singularities; a more detailed analysis will be postponed to a later section of these notes. Note finally that $\varphi(p_1 + \ldots + p_r) = \varphi(p_o + p_1 + \ldots + p_r)$, where p_o is the base point of the marked surface M, hence that $W_r \subseteq W_{r+1}$ for any r. There is consequently a chain of irreducible analytic subvarieties

$$M \cong W_1 \subset W_2 \subset \ldots \subset W_{g-1} \subset W_g = W_{g+1} = \ldots = J(M) ,$$

where dimension $W_r = r$ for $r = 1,\ldots,g$.

Viewed as a subset of the Picard variety $P(M) = \{\xi \in H^1(M, \mathcal{O}^*) | c(\xi) = 0\}$, note that the subvariety W_r consists precisely of those complex line bundles $\xi \in P(M)$ which can be written in the form $\xi = \zeta_{p_1} \cdots \zeta_{p_r}\zeta_{p_o}^{-r}$ for some points $p_1,\ldots,p_r$ on the surface M. This is of course in turn equivalent to the condition that the line bundle $\xi\zeta_{p_o}^{r}$ have a nontrivial holomorphic section; for $\zeta_{p_1}\cdots\zeta_{p_r}$ does have such a section, and conversely if $\xi\zeta_{p_o}^r$ has such a section with divisor $p_1 + \dots + p_r$ then $\xi\zeta_{p_o}^r = \zeta_{p_1}\cdots\zeta_{p_r}$. Writing $\gamma(\xi) = \dim_{\mathbb{C}} \Gamma(M, \mathcal{O}(\xi))$ for any complex line bundle ξ, it thus follows that

(1) $$W_r = \{\xi \in P(M) | \gamma(\xi\zeta_{p_o}^r) \geqq 1\} .$$

Incidentally, since $c(\xi\zeta_{p_o}^r) = r$ it is a familiar consequence of the Riemann-Roch theorem that $\gamma(\xi\zeta_{p_o}^r) \geqq 1$ whenever $r \geqq g$, and hence that $\xi \in W_r$ whenever $r \geqq g$, for any $\xi \in P(M)$; this then provides an alternative proof of the Jacobi inversion theorem, that $W_r = P(M)$ whenever $r \geqq g$.

An obvious and useful question to ask is what effect the natural algebraic operations of the group $J(M)$ have on this chain of irreducible analytic subvarieties. For any subset $S \subseteq J(M)$ it is possible to introduce the inverse set $-S = \{-s | s \in S\}$, and the translate $S + u = \{s+u | s \in S\}$; and for any pair of subsets $S, T \subseteq J(M)$ it is possible to introduce the sum $S + T = \{s+t | s \in S, t \in T\}$ and the quotient $S \ominus T = \{u \in J(M) | T + u \subseteq S\}$. This last set must of course be distinguished from the difference $S - T = S + (-T)$; the construction is a familiar one, in the context of ideals. If S is an analytic subvariety of $J(M)$ then so are $-S$ and $S + u$, since they arise from S by the application of an analytic homeomorphism of the manifold $J(M)$. If S and T are analytic subvarieties of $J(M)$ then as a consequence of the proper mapping theorem so is $S+T$, since it can be viewed as the image of the analytic mapping from the compact analytic variety $S \times T$ into $J(M)$ defined by $(s,t) \longrightarrow s+t$; and if S is an analytic subvariety so also is $S \ominus T$ for any subset T , since

$$(2) \qquad S \ominus T = \bigcap_{t \in T} (S - t)$$

is an intersection of analytic subvarieties of $J(M)$. These operations can be applied to the analytic subvarieties W_r , with the following results; for simplicity define $W_0 = 0$, the identity element of the group $J(M)$, noting that as a subset of the Picard variety $P(M)$ this set can be described as $W_0 = \{\xi \in P(M) | \gamma(\xi) \geqq 1\}$, paralleling the alternative description given for the other subvarieties W_r .

<u>Lemma 1.</u> For any integers $r,s \geqq 0$,

$$(3) \qquad W_r + W_s = W_{r+s} \ ;$$

and for any integers r,s such that $0 \leqq r \leqq s \leqq g-1$,

$$(4) \qquad W_s \ominus W_r = W_{s-r} \ .$$

Proof. The first assertion is a trivial consequence of the definition. As for the second assertion, viewing these sets as subvarieties of the Picard variety $P(M)$, note that

$$W_s \ominus W_r = \{\xi \in P(M) \mid \xi\eta \in W_s \quad \text{for all} \quad \eta \in W_r\}$$

$$= \{\xi \in P(M) \mid \gamma(\xi\eta\zeta_{p_o}^{s}) \geqq 1 \quad \text{for all} \quad \eta \in W_r\}$$

$$= \{\xi \in P(M) \mid \gamma(\xi\zeta_{p_1}\cdots\zeta_{p_r}\zeta_{p_o}^{s-r}) \geqq 1 \quad \text{for all} \quad p_1,\ldots,p_r \in M\}.$$

Now for any line bundle $\zeta \in H^1(M, \mathcal{O}^*)$ such that $0 \leqq c(\zeta) < g-1$ it follows easily though that $\gamma(\zeta\zeta_p) \geqq 1$ for all points $p \in M$ precisely when $\gamma(\zeta) \geqq 1$; and by iterating this observation it then follows that

$$W_s \ominus W_r = \{\xi \in P(M) \mid \gamma(\xi\zeta_{p_o}^{s-r}) \geqq 1\} = W_{s-r}$$

as desired. To complete the proof by demonstrating this auxiliary statement, note firstly that $\gamma(\zeta) \geqq 1$ obviously implies that $\gamma(\zeta\zeta_p) \geqq 1$ for all points $p \in M$. On the other hand if $\gamma(\zeta) = 0$ it follows from the Riemann-Roch theorem that $\gamma(\kappa\zeta^{-1}) = g-1-c(\zeta) > 0$ and that $\gamma(\kappa\zeta^{-1}\zeta_p^{-1}) = g-2-c(\zeta) + \gamma(\zeta\zeta_p)$, where $\kappa \in H^1(M, \mathcal{O}^*)$ is the canonical bundle. Now selecting a point $p \in M$ at which at least one of the holomorphic sections of the bundle $\kappa\zeta^{-1}$ is non-zero, there are evidently at most $g-2-c(\zeta)$ linearly independent holomorphic sections of $\kappa\zeta^{-1}$ which vanish at p ; but the dimensions of the space of holomorphic sections of $\kappa\zeta^{-1}$ which vanish at p is $\gamma(\kappa\zeta^{-1}\zeta_p^{-1}) = g-2-c(\zeta) + \gamma(\zeta\zeta_p)$, hence $\gamma(\zeta\zeta_p) = 0$ at that point, and the proof is concluded.

The condition that $s \leqq g-1$ in the second assertion of the lemma is of course quite necessary, since for $s \geqq g$ necessarily $W_s = J(M)$ and consequently $W_s \ominus W_r = J(M)$ as well. Note that having proved Lemma 1, the corresponding statements for translates of these subvarieties follow quite trivially; hence

$$(5) \qquad (W_r + u) + (W_s + v) = W_{r+s} + (u+v)$$

for any integers $r, s \geqq 0$ and any points $u, v \in J(M)$, and

$$(6) \qquad (W_s + v) \ominus (W_r + u) = W_{s-r} + (v-u)$$

for any integers $0 \leqq r \leqq s \leqq g-1$ and any points $u, v \in J(M)$. A special case of the lemma which merits particular mention is the assertion that <u>if</u> $W_r + u = W_r$ <u>for some integer</u> $0 \leqq r \leqq g-1$ <u>and some point</u> $u \in J(M)$, <u>then necessarily</u> $u = 0$, since $u \in W_r \ominus W_r = W_0$; thus the subsets $W_r \subseteq J(M)$ are quite far from being preserved by translations. Another special case which also merits note is the assertion that

$$(7) \qquad W_1 = W_{g-1} \ominus W_{g-2} = \bigcap_{u \in W_{g-2}} (W_{g-1} - u) ;$$

this equation shows that the submanifold $W_1 \cong M$ can be recovered merely from the terminal portion $W_{g-2} \subset W_{g-1} \subset J(M)$ of this sequence of subvarieties, hence that the original Riemann surface is determined by that portion of the sequence.

(c) When viewed as a subset of the Picard variety, the subvariety of positive divisors was characterized as $W_r = \{\xi \in P(M) | \gamma(\xi\zeta_{p_o}^r) \geqq 1\}$; and from this point of view it is only natural to introduce the further subsets

(8) $$W_r^\nu = \{\xi \in P(M) | \gamma(\xi\zeta_{p_o}^r) \geqq \nu\}$$

for arbitrary integers $\nu \geqq 1$. These subsets will be called the subsets of <u>special positive divisors</u> for $\nu > 1$, and form a descending filtration

$$W_r = W_r^1 \supseteq W_r^2 \supseteq W_r^3 \supseteq \cdots$$

of the subvariety W_r of positive divisors of degree r . They can also be characterized quite conveniently as follows.

<u>Lemma 2.</u> For any integers $r \geqq 0$ and $\nu \geqq 1$,

(9) $$W_r^\nu = W_{r-\nu+1} \ominus (-W_{\nu-1}) \quad \text{whenever} \quad \nu \leqq r+1 \text{ , and}$$

(10) $$W_r^\nu = \emptyset \quad \text{whenever} \quad \nu > r+1 \text{ .}$$

Proof. Letting $x \in J(M)$ correspond to a line bundle $\xi \in P(M)$, it follows from the definition that $x \in W_r^\nu$ precisely when $\gamma(\xi\zeta_{p_o}^r) \geqq \nu$. Now the existence of at least ν linearly independent holomorphic sections of the bundle $\xi\zeta_{p_o}^r$ is clearly equivalent to the condition that there exists a nontrivial holomorphic section of the bundle $\xi\zeta_{p_o}^r$ with zeros at an arbitrarily specified set of $\nu-1$ points of the Riemann surface, hence to the condition that $\gamma(\xi\zeta_{p_o}^{r}\zeta_{p_1}^{-1}\cdots\zeta_{p_{\nu-1}}^{-1}) \geqq 1$ for arbitrary points $p_1,\ldots,p_{\nu-1}$ on the Riemann surface. Since this line bundle has Chern class $r-\nu+1$, the last condition is evidently impossible whenever $r-\nu+1 < 0$, so that $W_r^\nu = \emptyset$ in that case. Otherwise the last condition can be rewritten in the form $\gamma(\xi\zeta_{p_o}^{\nu-1}\zeta_{p_1}^{-1}\cdots\zeta_{p_{\nu-1}}^{-1}\cdot\zeta_{p_o}^{r-\nu+1}) \geqq 0$, and is thus in turn equivalent to the condition that $\xi\zeta_{p_o}^{\nu-1}\zeta_{p_1}^{-1}\cdots\zeta_{p_{\nu-1}}^{-1} \in W_{r-\nu+1} \subseteq P(M)$, or in terms of the Jacobi variety, that $x - (p_1 + \cdots + p_{\nu-1}) \in W_{r-\nu+1}$, for any points $p_1,\ldots,p_{\nu-1}$ on the Riemann surface; but this is precisely the definition of the subset $W_{r-\nu+1} \ominus (-W_{\nu-1})$, and the proof is thereby concluded.

It is an immediate consequence of this lemma that the subsets W_r^ν are complex analytic subvarieties of the Jacobi variety $J(M)$, indeed that they can be written in the form

$$(11)\qquad W_r^\nu = \bigcap_{u \in W_{\nu-1}} (W_{r-\nu+1} + u)$$

whenever $\nu \leqq r+1$, and are otherwise empty. If the notation is extended by putting $W_r = \emptyset$ whenever $r < 0$, then equation (10) can be subsumed as a special case of equation (9). The assertions of the lemma can also be inverted, yielding the equation

$$(12)\qquad W_s \ominus (-W_r) = W_{r+s}^{r+1}$$

for any integers $r \geqq 0$, $s \geqq 0$.

If $r > 2g-2$ it follows immediately from the Riemann-Roch theorem that $\gamma(\xi\zeta_{p_o}^{r}) = r-g+1$, so that $W_r^\nu = \emptyset$ when $\nu > r-g+1$ and $W_r^\nu = J(M)$ when $\nu \leqq r-g+1$; and from Lemma 2 it follows that $W_r^\nu = \emptyset$ when $\nu > r+1$, and that $W_r^\nu = \emptyset$ when $\nu = r+1 > 2$ and $W_r^\nu = 0$ when $\nu = r+1 = 2$. These cases being quite trivial, it is evidently only reasonable to restrict attention to those subvarieties W_r^ν with indices in the range $1 \leqq \nu \leqq r \leqq 2g-2$.

As a useful bit of further notation, note that the divisors of any two holomorphic or meromorphic Abelian differentials are linearly equivalent, hence that the image $k = \varphi(\mathfrak{z}) \in J(M)$ of any such divisor $\mathfrak{z}$ is a well defined point of the Jacobi variety; this point will be called the canonical point of $J(M)$. In terms of the Picard variety, the canonical point k is represented by the line bundle $\kappa\zeta_{p_o}^{2-2g} \in P(M)$, where κ is the canonical bundle of the Riemann surface. It follows from the Riemann-Roch theorem that

$$\gamma(\kappa\zeta_{p_o}^{2-2g}\xi^{-1}\cdot\zeta_{p_o}^{2g-r-2}) = \gamma(\kappa\cdot\xi^{-1}\zeta_{p_o}^{-r}) = \gamma(\xi\zeta_{p_o}^{r}) + g-r-1$$

for any line bundle $\xi \in P(M)$; note that if $\xi \in P(M)$ represents a point $x \in J(M)$, then $\kappa\zeta_{p_o}^{2-2g}\xi^{-1} \in P(M)$ represents the point $k-x \in J(M)$. Now $x \in W_r^\nu$ precisely when $\gamma(\xi\zeta_{p_o}^{r}) \geqq \nu$, hence precisely when $\gamma(\kappa\zeta_{p_o}^{2-2g}\xi^{-1}\cdot\zeta_{p_o}^{2g-r-2}) \geqq \nu+g-r-1$; and if $2g-r-2 \geqq 0$ and $\nu+g-r-1 \geqq 1$, this last inequality is just the condition that $k-x \in W_{2g-r-2}^{\nu+g-r-1}$. Thus the Riemann-Roch theorem leads immediately to the identity

$$(13) \qquad k - W_r^\nu = W_s^\mu$$

where $s = 2g-2-r$ and $\mu = g-1 - (r-\nu)$, provided of course that $r, s \geqq 0$ and $\nu, \mu \geqq 1$ so that both sides are well defined. Note first of all that when $g-1 \leqq r \leqq 2g-2$ then necessarily $0 \leqq s \leqq g-1$; thus when considering the analytic properties of the separate subvarieties W_r^ν , rather than their locations and interrelations in $J(M)$, <u>it is really sufficient to restrict attention to those subvarieties with indices in the range</u> $1 \leqq \nu \leqq r \leqq g-1$. The most symmetric case of formula (13) is that in which $r = s = g-1$, and in this case

$$(14) \qquad k - W_{g-1}^\nu = W_{g-1}^\nu$$

for $1 \leqq \nu \leqq g-1$. In particular, when $\nu = 1$, it follows that $k - W_{g-1} = W_{g-1}$, hence that $k = W_{g-1} \ominus (-W_{g-1})$; and this condition of course determines the canonical point uniquely.

Now Lemma 2 asserts that $W_r^\nu = W_{r-\nu+1} \ominus (-W_{\nu-1})$ whenever $1 \leqq \nu \leqq r$, and if in addition $r-\nu+1 \leqq g$ and $\nu-1 \leqq g$ then $W_{r-\nu+1}$ and $-W_{\nu-1}$ are analytic subvarieties of $J(M)$ of dimensions $r-\nu+1$ and $\nu-1$ respectively; and clearly then $W_r^\nu = \emptyset$ whenever $\nu-1 > r-\nu+1$, since in that case no translate of $-W_{\nu-1}$ can possibly be contained in $W_{r-\nu+1}$. Thus

$$(15) \qquad W_r^\nu = \emptyset \quad \text{whenever} \quad 2\nu > r+2$$

where $1 \leqq \nu \leqq r \leqq 2g-2$ and $\nu \leqq g+1$. Furthermore, in the borderline case that $2\nu = r+2$, the subvariety $W^{\nu}_{2\nu-2} = W_{\nu-1} \ominus (-W_{\nu-1})$ consists of those points $u \in J(M)$ such that $u - W_{\nu-1} \subseteq W_{\nu-1}$, or equivalently such that $-W_{\nu-1} = W_{\nu-1} - u$ since $W_{\nu-1}$ and its translates are irreducible subvarieties; and there is obviously at most one such point $u \in J(M)$. Thus

(16) either $W^{\nu}_{2\nu-2} = \emptyset$ or $W^{\nu}_{2\nu-2}$ consists of the unique point $u \in J(M)$ such that $-W_{\nu-1} = W_{\nu-1} - u$, whenever $2 \leqq \nu \leqq g$.

Of course in the special case $\nu = g$, the set $W^{g}_{2g-2} = W_{g-1} \ominus (-W_{g-1}) = k$ consists precisely of the canonical point, as noted earlier.

Note that (15) can be rephrased as the inequality $\gamma(\xi\zeta^{r}_{p_o}) < \nu$ for all points $\xi \in P(M)$, whenever $2\nu > r+2$ and $1 \leqq \nu \leqq r \leqq 2g-2$, $\nu \leqq g+1$; the bundles $\xi\zeta^{r}_{p_o}$ are of course all the line bundles of Chern class r . This is easily seen to be equivalent to the inequality

(17) $\gamma(\xi) \leqq [\frac{1}{2} c(\xi)] + 1$ for all line bundles $\xi \in H^1(M, \mathcal{O}^*)$ with Chern classes in the range $3 \leqq c(\xi) \leqq 2g-2$,

where as usual $[x]$ denotes the greatest integer function. This represents a significant improvement in the estimate of the maximal values for $\gamma(\xi)$ derived in the earlier lecture notes; the table on page 113 of those notes can thus be replaced the the following table of inequalities relating the Chern class $c(\xi)$ and the dimension $\gamma(\xi) = \dim_{\mathbf{C}} \Gamma(M, \mathcal{O}(\xi))$ for any line bundle $\xi \in H^1(M, \mathcal{O}^*)$.

(18)

$c(\xi)$:	0	1	2	3	4	5	6	7	...	$g-1$	g	$g+1$	...
max $\gamma(\xi)$:	1	2	2	2	3	3	4	4	...	$[\frac{g-1}{2}]+1$	$[\frac{g}{2}]+1$	$[\frac{g+1}{2}]+1$	...
min $\gamma(\xi)$:	0	0	0	0	0	0	0	0	...	0	1	2	...

$c(\xi)$:	$2g-5$	$2g-4$	$2g-3$	$2g-2$	$2g-1$	$2g$	...
max $\gamma(\xi)$:	$g-2$	$g-1$	$g-1$	g	g	$g+1$	...
min $\gamma(\xi)$:	$g-4$	$g-3$	$g-2$	$g-1$	g	$g+1$	...

Note by the way that if ξ is a line bundle with even Chern class $c(\xi) = 2\nu \leqq 2g-2$ and with the maximal number of holomorphic sections as listed in this table $\gamma(\xi) = \nu+1$, then $\xi\zeta_{p_0}^{-2\nu} \in P(M)$ represents a point of the subvariety $W_{2\nu}^{\nu+1}$; but as noted in (16) this subvariety is either empty or consists of but a single point. Thus on any Riemann surface there can exist at most one line bundle ξ with Chern class $c(\xi) = 2\nu \leqq 2g-2$ such that this maximum $\gamma(\xi) = \nu+1$ is actually attained. It will appear later that there are Riemann surfaces (indeed the hyperelliptic Riemann surfaces) for which this maximum value is attained; and of course tensoring such a bundle with an arbitrary point bundle will yield a line bundle of odd Chern class for which the maximum is also attained. Thus the estimates provided by this table (18) or by the inequality (17) are the best possible, in general.

Consider next a fixed point $p \in M$ with image $\varphi(p) = x \in W_1 \subseteq J(M)$, represented by the line bundle $\xi = \zeta_p\zeta_{p_0}^{-1} \in P(M)$; and note that for any integer $r \geqq 1$ the divisor $r\cdot p$ has image $\varphi(r\cdot p) = r\cdot x \in r\cdot W_1 \subseteq W_r \subseteq J(M)$, represented by the line bundle $\xi^r \in P(M)$. The subvariety $r\cdot W_1 \subseteq W_r$ thus corresponds to the subset of positive divisors that can be represented as based at a single point on the Riemann surface M; and these divisors were considered in some detail in the earlier lecture notes. In particular, in Theorem 14 of those lecture notes it was demonstrated that $r+1 - \gamma(\zeta_p^r)$ is equal to the number of Weierstrass gap values at the point p in the sequence $1,2,\ldots,r$; that is of course equivalent to the assertion that $\gamma(\zeta_p^r) -1$ is equal to the number of nongaps at the point p occurring in the sequence $1,2,\ldots,r$. Now $r\cdot x \in W_r^\nu$ precisely when $\gamma(\zeta_p^r) = \gamma(\xi^r\zeta_{p_0}^r) \geqq \nu$, hence using the preceding observation, precisely when there are at least $\nu-1$ nongaps at the point p among the sequence $1,2,\ldots,r$; thus

$$(19) \qquad r\cdot W_1 \cap W_r^\nu = \{r\cdot\varphi(p) \mid \text{there are at least } \nu-1 \text{ nongaps at } p \text{ in } (1,2,\ldots,r)\}.$$

At a point $p \in M$ which is not a Weierstrass point the gap sequence is precisely $1,2,\ldots,g$; hence for $1 \leqq \nu \leqq r \leqq g$ the intersection $r\cdot W_1 \cap W_r^\nu$ can only consist of multiples of Weierstrass points on W_1, so is necessarily a finite set of points. Some at least of these intersections are necessarily nonempty, depending on the gap structure at the Weierstrass points.

Turning now to more general properties of these subvarieties of special positive divisors, Lemma 2 would seem to indicate that they should be particularly well behaved under the operation $\ominus$, and that is indeed the case.

Lemma 3. For any integers $1 \leqq \nu \leqq r$, $0 \leqq s$,

(20) $$W_r^\nu \ominus (-W_s) = W_{r+s}^{\nu+s} ,$$

(21) $$W_r^\nu \ominus W_s = W_{r-s}^\nu \quad \text{provided} \quad s \leqq r-\nu , \text{ and}$$

(22) $$W_r^\nu \ominus (W_s - W_s) = W_r^{\nu+s} \quad \text{provided} \quad s \leqq r-\nu .$$

Proof. These assertions are immediate consequences of Lemmas 1 and 2 and of the elementary observation that $(A \ominus B) \ominus C = A \ominus (B + C)$ for any subsets A, B, C of $J(M)$. To verify this observation, note that $x \in (A \ominus B) \ominus C$ precisely when $x + c \in A \ominus B$ for all $c \in C$, hence precisely when $x + c + b \in A$ for all $b \in B$ and $c \in C$; and the last condition is just that $x \in A \ominus (B + C)$. Now to prove (20) note that
$W_r^\nu \ominus (-W_s) = [W_{r-\nu+1} \ominus (-W_{\nu-1})] \ominus (-W_s) = W_{r-\nu+1} \ominus (-W_{\nu-1} - W_s) =$
$= W_{r-\nu+1} \ominus (-W_{s+\nu-1}) = W_{r+s}^{\nu+s}$; to prove (21) note that
$W_r^\nu \ominus W_s = [W_{r-\nu+1} \ominus (-W_{\nu-1})] \ominus W_s = W_{r-\nu+1} \ominus (W_s - W_{\nu-1}) =$
$= [W_{r-\nu+1} \ominus W_s] \ominus (-W_{\nu-1}) = W_{r-s-\nu+1} \ominus (-W_{\nu-1}) = W_{r-s}^\nu$; and to prove (22) note that as a consequence of (20) and (21) it follows that
$W_r^\nu \ominus (W_s - W_s) = [W_r^\nu \ominus W_s] \ominus (-W_s) = W_{r-s}^\nu \ominus (-W_s) = W_r^{\nu+s}$, thus completing the proof.

Although the subsets W_r^ν are analytic subvarieties of the Jacobi variety, they may not be irreducible subvarieties if $\nu > 1$. It is still possible to speak of the dimensions of these subvarieties though, where as usual the dimension of an analytic subvariety is defined to be the maximum of the dimensions of its irreducible components; this dimension will be denoted $\dim W_r^\nu$. Recall that $\dim W_r^1 = \dim W_r = r$ for $1 \leqq r \leqq g$, while $\dim W_r^1 = \dim W_r = g$ for $g \geqq r$.

Lemma 4. If the subvariety W_r^ν is nonempty, then

(23) $\dim W_{r+1}^{\nu+1} < \dim W_r^\nu$ whenever $1 \leqq \nu \leqq r \leqq g-1$, and

(24) $\dim W_{r-1}^{\nu} < \dim W_r^\nu$ whenever $1 \leqq \nu < r \leqq g-1$.

Proof. Note that $W_r^\nu \subseteq W_{g-1}$ whenever $r \leqq g-1$, so that W_r^ν is then necessarily a proper analytic subvariety of $J(M)$. If $W_{r+1}^{\nu+1} = \emptyset$ the first assertion of the lemma is trivial. Otherwise select an irreducible component V of the subvariety $W_{r+1}^{\nu+1}$ with $\dim V = \dim W_{r+1}^{\nu+1}$; and note that $V \subseteq V - W_1 \subseteq W_{r-1}^{\nu+1} - W_1 \subseteq W_r^\nu$, since $W_{r+1}^{\nu+1} = W_r^\nu \ominus (-W_1)$ by Lemma 3. Now $V - W_1$ is an irreducible analytic subvariety of $J(M)$, since it is the image of the irreducible variety $V \times W_1$ under the obvious analytic mapping $V \times W_1 \longrightarrow J(M)$; and since $V \subseteq V - W_1$, either $V = V - W_1$ or $\dim V < \dim(V - W_1)$. If $V = V - W_1$, then $V = V - W_1 = V - W_2 = \dots = V - W_g = J(M)$, which is impossible since V is contained in W_r^ν and W_r^ν is a proper analytic subvariety of $J(M)$; therefore $\dim W_{r+1}^{\nu+1} = \dim V < \dim(V - W_1) \leqq \dim W_r^\nu$, which is the first assertion of the lemma. The proof of the second assertion is quite similar. If $W^{\nu}_{r-1} \neq \emptyset$, select an irreducible component V such that $\dim V = \dim W^{\nu}_{r-1}$, and note that $V \subseteq V + W_1 \subseteq W^{\nu}_{r-1} + W_1 \subseteq W^{\nu}_{r}$ since $W^{\nu}_{r-1} = W^{\nu}_{r} \ominus W_1$ by Lemma 3; again $V + W_1$ is irreducible, and it is impossible that $V = V + W_1$, so necessarily $\dim W^{\nu}_{r-1} = \dim V < \dim(V + W_1) \leq \dim W^{\nu}_{r}$. That suffices to conclude the proof.

<u>Theorem 7.</u> The subvarieties W^{ν}_{r} of special positive divisors on a Riemann surface are analytic subvarieties of the Jacobi variety such that

$$\dim W^{\nu}_{r} \leq r - 2\nu + 2 \quad \text{for} \quad 2 \leq \nu \leq r \leq g-1 .$$

Proof. By successive application of the inequality (23) of Lemma 4 it follows that

$$\dim W^{\nu}_{r} < \dim W^{\nu-1}_{r-1} < \dots < \dim W^{1}_{r-(\nu-1)} .$$

Each of these $\nu-1$ inequalities must cut the dimension by at least 1, so that $\dim W^{\nu}_{r} \leq \dim W^{1}_{r-(\nu-1)} - (\nu-1)$; and since $\dim W^{1}_{r-\nu+1} = \dim W_{r-\nu+1} = r - \nu + 1$, it follows immediately that $\dim W^{\nu}_{r} \leq r - 2\nu + 2$, as desired.

(d) This maximum is actually attained for hyperelliptic Riemann surfaces, which are particularly rich in functions and are exceptional in several ways. Recall that a hyperelliptic Riemann surface is one that can be represented as a two sheeted branched analytic covering of the projective line; and that each such surface M has a unique analytic automorphism $\theta: M \longrightarrow M$ of period 2 corresponding to the interchange of sheets in any representation of M as a two sheeted branched analytic covering of the projective line. The meromorphic functions on M invariant under this automorphism can be identified with the meromorphic functions on the projective line; hence for any two points p, q of M there will exist a meromorphic function having divisor $p + \theta p - q - \theta q$. Thus any two divisors on M of the form $p + \theta p$ are linearly equivalent; the common image $e = \varphi(p + \theta p) \in J(M)$ of all such divisors is a well defined point of the Jacobi variety, which will be called the <u>hyperelliptic point</u> of $J(M)$. In terms of the Picard variety, the hyperelliptic point is represented by any line bundle of the form $\zeta_p \zeta_{\theta p} \zeta_{p_o}^{-2}$. Such a line bundle $\eta = \zeta_p \zeta_{\theta p} \zeta_{p_o}^{-2}$ has the property that $\gamma(\eta \zeta_{p_o}^2) = 2$, so that $e \in W_2^2$; and conversly, whenever W_2^2 is nonempty its unique point is represented by a line bundle η such that $c(\eta \zeta_{p_o}^2) = 2$ and $\gamma(\eta \zeta_{p_o}^2) = 2$, hence the surface is hyperelliptic and the line bundle η represents the hyperelliptic point. Thus hyperelliptic Riemann surfaces can be characterized as those for which $W_2^2 \neq \emptyset$; and the unique point in W_2^2 is then the hyperelliptic point.

Now if e is the hyperelliptic point on the Jacobi variety of a hyperelliptic Riemann surface it follows from (16) that $-W_1 = W_1 - e$; and iterating this relation,

$$-W_{\nu-1} = - W_1 - \dots - W_1 = (W_1 - e) + \dots + (W_1 - e) = W_{\nu-1} - (\nu-1)\cdot e .$$

Thus $W^{\nu}_{2\nu-2}$ is nonempty, indeed consists of the point $(\nu-1)\cdot e$, whenever $2 \leqq \nu \leqq g$. In particular, the canonical point is given in terms of the hyperelliptic point by the formula $k = (g-1)\cdot e$. It further follows, as already noted as a consequence of the condition $W^{2}_{2\nu-2} \neq \emptyset$, that the maximum values for $\gamma(\xi)$ given by formula (17) on the table (18) are actually attained on hyperelliptic Riemann surfaces. Finally note also that $W^{\nu}_{r} = W_{r-\nu+1} \ominus (-W_{\nu-1}) = W_{r-\nu+1} \ominus (W_{\nu-1} - (\nu-1)\cdot e) = (W_{r-\nu+1} \ominus W_{\nu-1}) - (\nu-1)\cdot e = W_{r-2\nu+2} - (\nu-1)\cdot e$ whenever $1 \leqq \nu \leqq r \leqq g-1$; thus <u>on a hyperelliptic Riemann surface</u>

$$W^{\nu}_{r} = W_{r-2\nu+2} - (\nu-1)\cdot e \tag{25}$$

whenever $1 \leqq \nu \leqq r \leqq g-1$, so the subvarieties of special position divisors are irreducible analytic subvarieties of the maximal possible dimension in this case.

The hyperelliptic Riemann surfaces not only provide examples of surfaces for which the maximal values described in the preceding discussion are actually attained, but are even characterized by the attainment of these maximal values. A partial result in this direction is the following.

<u>Theorem 8 (Clifford's Theorem).</u> If $W^{\nu}_{2\nu-2} \neq \emptyset$ for some index ν in the range $2 \leqq \nu \leqq g-1$, for a Riemann surface of genus g, then that surface is hyperelliptic.

Proof. As noted earlier, hyperelliptic Riemann surfaces are characterized by the condition that $W_2^2 \neq \emptyset$; so in order to prove the theorem it is sufficient to show that if $W_{2\nu-2}^{\nu} \neq \emptyset$ for some index ν in the range $2 < \nu \leqq g-1$, then $W_{2\lambda-2}^{\lambda} \neq \emptyset$ for some index λ in the range $2 \leqq \lambda < \nu$. Note first of all that from the Riemann-Roch theorem in the form (13) it follows that $k - W_{2\nu-2}^{\nu} = W_{2\mu-2}^{\mu}$ where $\mu + \nu = g + 1$; and it can of course be assumed that $\nu \leqq \mu$. (Note that this is the point at which the value $\nu = g$ must be excluded, since then $\mu = 1$.) Assuming that $W_{2\nu-2}^{\nu} \neq \emptyset$, choose a point $x \in W_{2\nu-2}^{\nu} \subseteq J(M)$ represented by a line bundle $\xi \in P(M)$; and let $y = k - x \in W_{2\mu-2}^{\mu} \subseteq J(M)$ be represented by a line bundle $\eta \in P(M)$, so that $\xi\eta\zeta_{p_0}^{2g-2} = \kappa$ is the canonical bundle. The point x can be described as the image under the Jacobi homomorphism of a divisor $\mathscr{J}_x$ of degree $2\nu-2$ on the surface M , and correspondingly y can be described as the image of a divisor $\mathscr{J}_y$ of degree $2\mu-2$. Since $x \in W_{2\nu-2}^{\nu}$ with $\nu \geqq 3$, two points of the divisor $\mathscr{J}_x$ can be specified quite arbitrarily, so it can be assumed that at least one point of $\mathscr{J}_x$ also appears in $\mathscr{J}_y$ and that at least one point of $\mathscr{J}_x$ does not appear in $\mathscr{J}_y$; the set of common points (counting multiplicities) of the two divisors $\mathscr{J}_x$ and $\mathscr{J}_y$ can be viewed as forming a divisor $\mathscr{J}_z$ of degree r , where $1 \leqq r < 2\nu-2$ as a consequence of these choices, and that divisor has as image under the Jacobi homeomorphism a point $z \in W_r \subseteq P(M)$. Now recall that $\gamma(\xi\zeta_{p_0}^{2\nu-2})$ can also be described as the dimension of the complex vector space

$L(\mathscr{D}_x)$ consisting of those meromorphic functions f on the Riemann surface M such that $\mathscr{D}(f) + \mathscr{D}_x \geqq 0$, and correspondingly $\gamma(\eta\zeta_{p_o}^{2\mu-2})$ is the dimension of the complex vector space $L(\mathscr{D}_y)$ associated to the divisor $\mathscr{D}_y$, as described on page 57 and the following pages in the earlier lecture notes. Clearly $L(\mathscr{D}_x) \cap L(\mathscr{D}_y) = L(\mathscr{D}_z)$, so that

$$\dim_{\mathbf{C}}[L(\mathscr{D}_x) + L(\mathscr{D}_y)] = \dim_{\mathbf{C}} L(\mathscr{D}_x) + \dim_{\mathbf{C}} L(\mathscr{D}_y) - \dim_{\mathbf{C}} L(\mathscr{D}_z) ;$$

and since it is also evident that

$$L(\mathscr{D}_x) + L(\mathscr{D}_y) \subseteqq L(\mathscr{D}_x + \mathscr{D}_y - \mathscr{D}_z)$$

it follows that

$$\dim_{\mathbf{C}}[L(\mathscr{D}_x) + L(\mathscr{D}_y)] \leqq \dim_{\mathbf{C}} L(\mathscr{D}_x + \mathscr{D}_y - \mathscr{D}_z) .$$

Rewriting these observations in terms of the dimensions of the spaces of holomorphic sections of the appropriate line bundles, it then follows that

$$\gamma(\xi\zeta_{p_o}^{2\nu-2}) + \gamma(\eta\zeta_{p_o}^{2\mu-2}) - \gamma(\zeta\zeta_{p_o}^{r}) \leqq \gamma(\xi\eta\zeta^{-1}\zeta_{p_o}^{2g-2-r})$$

where $\zeta \in P(M)$ is the line bundle corresponding to the point $z \in J(M)$; and since $\eta\zeta_{p_o}^{2\mu-2} = \kappa(\xi\zeta_{p_o}^{2\nu-2})^{-1}$ and $\xi\eta\zeta^{-1}\zeta_{p_o}^{2g-2-r} = \kappa(\zeta\zeta_{p_o}^{r})^{-1}$, the Riemann-Roch theorem can be applied once again to rewrite this inequality in the form

$$2\gamma(\xi\zeta_{p_o}^{2\nu-2}) - (2\nu-2) \leqq 2\gamma(\zeta\zeta_{p_o}^{r}) - r .$$

Since $x \in W_{2\nu-2}^{\nu}$, the left hand side in this last inequality is

$\geqq 2\nu - (2\nu-2) = 2$; and from the upper bound provided by (17), the right hand side in this inequality is $\leqq 2([\frac{r}{2}]+1) - r$. It thus follows that

$$2 \leqq 2\gamma(\zeta\zeta_{p_o}^{r}) - r \leqq 2[\tfrac{r}{2}] + 2 - r \; ;$$

this can only happen when r is an even integer, say $r = 2\lambda-2$, and when $\gamma(\zeta\zeta_{p_o}^{r}) = \lambda$. This in turn means that $z \in W_{2\lambda-2}^{\lambda}$, hence that $W_{2\lambda-2}^{\lambda} \neq \emptyset$ for some index λ in the range $2 \leqq \lambda < \nu-1$; and as already noted, that suffices to conclude the proof of the theorem.

An immediate corollary of this theorem is the observation that the inequality (17) can be improved on a Riemann surface that is not hyperelliptic; for any line bundle $\xi \in H^1(M, \mathcal{O}^*)$ with even Chern class $c(\xi) = 2\nu-2$ it follows that $\gamma(\xi) \leqq [\frac{1}{2}\, c(\xi)]$ whenever $3 \leqq c(\xi) \leqq 2g-2$, since a line bundle with $\gamma(\xi) = [\frac{1}{2}\, c(\xi)] + 1 = \nu$ represents a point in $W_{2\nu-2}^{\nu}$. Thus

(26) $\gamma(\xi) \leqq [\frac{c(\xi)+1}{2}]$ for all line bundles $\xi \in H^1(M, \mathcal{O}^*)$ with Chern classes in the range $3 \leqq c(\xi) \leqq 2g-3$, when the Riemann surface is <u>not</u> hyperelliptic.

Correspondingly, when M is <u>not</u> hyperelliptic the table of inequalities (18) can be improved also, as follows.

(27)

$c(\xi)$:	0	1	2	3	4	5	6	7	...	g	$g+1$
$\max\gamma(\xi)$:	1	2	2	2	2	3	3	4	...	$[\frac{g+1}{2}]$	$[\frac{g+2}{2}]$
$\min\gamma(\xi)$:	0	0	0	0	0	0	0	0	...	1	2

$c(\xi)$:	...	$2g-5$	$2g-4$	$2g-3$	$2g-2$	$2g-1$	$2g$	...
$\max\gamma(\xi)$:	...	$g-2$	$g-2$	$g-1$	g	g	$g+1$	...
$\min\gamma(\xi)$:	...	$g-4$	$g-3$	$g-2$	$g-1$	g	$g+1$	...

Notes for §2.

(a) This section is really just a review, although from a slightly different point of view, of material that can be found in §8 of the earlier lecture notes on Riemann surfaces. For most if not all of the results in §2, it is not really necessary to select a particular coordinate representation of the Picard variety $P(M)$, such as the representation given here in terms of the Jacobi variety; indeed, the discussion in the remainder of §2 can be rephrased entirely in terms of the Picard variety, continuing more directly the treatment begun in §8 of the earlier lecture notes. However there are some later points at which the explicit representation of the Picard variety is useful, if not essential, particularly when discussing homotopy rather than homology invariants; and the change in point of view may serve as a useful element in reviewing the relevant portions of the earlier lecture notes.

The Jacobi mapping $\varphi: M \longrightarrow J(M)$ has a useful functorial property, which can conveniently be described as follows. Consider an arbitrary compact complex analytic torus $J^* = \mathbf{C}^n/\mathfrak{L}^*$, defined by a lattice subgroup $\mathfrak{L}^* \subset \mathbf{C}^n$, and a complex analytic mapping $\psi: M \longrightarrow J^*$. This mapping obviously lifts to a complex analytic mapping $\tilde{\psi}: \tilde{M} \longrightarrow \mathbf{C}^n$ from the universal covering surface $\tilde{M}$ of the Riemann surface M into the universal covering space $\mathbf{C}^n$ of the torus J^*; and the component functions of the lifted mapping are complex analytic functions $\tilde{\psi}_i(z)$ on $\tilde{M}$ such that for any $T \in \Gamma$ the values $\tilde{\psi}_i(Tz) - \tilde{\psi}_i(z)$ are constants, indeed are the components of a vector belonging to the lattice $\mathfrak{L}^*$. The functions $\tilde{\psi}_i(z)$ are consequently Abelian integrals on $\tilde{M}$, so can be expressed in terms of the canonical Abelian integrals in the form

$$\tilde{\psi}_i(z) = \sum_{j=1}^{g} c_{ij}w_j(z) + c_i \ , \qquad 1 \le i \le n \ ,$$

for some constants c_{ij} and c_i ; and it is clear that the constants c_{ij} must have the property that $\sum_{j=1}^{g} c_{ij}\omega_j(T)$, $1 \le i \le n$, are the components of a vector in the lattice $\mathfrak{L}^*$ for any transformation $T \in \Gamma$. The matrix $C = \{c_{ij}\}$ can now be viewed as determining a linear transformation $C\colon \mathbb{C}^g \longrightarrow \mathbb{C}^n$ such that $C(\mathfrak{L}) \subseteq \mathfrak{L}^*$; and this in turn induces a complex analytic group homomorphism $C\colon J(M) \longrightarrow J^*$. Conversely it is clear that any complex analytic group homomorphism $C\colon J(M) \longrightarrow J^*$ arises from a linear transformation $C\colon \mathbb{C}^g \longrightarrow \mathbb{C}^n$ such that $C(\mathfrak{L}) \subseteq \mathfrak{L}^*$. Thus altogether <u>the mapping</u> $\psi\colon M \longrightarrow J(M)$ <u>must be of the form</u> $\psi = C\varphi + c$, <u>where</u> $C\colon J(M) \longrightarrow J^*$ <u>is an arbitrary complex analytic group homomorphism,</u> c <u>is any point in the group</u> J^* , <u>and</u> $\varphi\colon M \longrightarrow J(M)$ <u>is the Jacobi mapping.</u> In particular, any complex analytic mapping $\psi\colon M \longrightarrow J(M)$ of a Riemann surface M into its Jacobi variety must be of the form $\psi = C\varphi + c$ for some endomorphism $C\colon J(M) \longrightarrow J(M)$ and some point $c \in J(M)$. It is fairly evident how this observation can be used to give a functorial characterization of the Jacobi variety, but details will be left to the reader; the compact complex torus defined in this functorial manner is often called the <u>Albanese variety</u> of the Riemann surface M.

The Jacobi homomorphism $\varphi\colon \Gamma(M, \mathcal{D}) \longrightarrow J(M)$ can of course be described correspondingly. Consider an arbitrary complex analytic mapping $\psi\colon M^r \longrightarrow J(M)$, for some index $r \geqq 1$. Note that for any fixed points $p_2,\ldots,p_r \in M$ the restriction $\psi(p,p_2,\ldots,p_r)$ can be viewed as a complex analytic mapping from M into $J(M)$, hence can be written in the form

$$\psi(p,p_2,\ldots,p_r) = C_1\varphi(p) + c_1$$

for some endomorphism $C_1\colon J(M) \longrightarrow J(M)$ and some point $c_1 \in J(M)$; and the matrix C_1 and the point c_1 evidently depend analytically on the points $p_2,\ldots,p_r \in M$. Now it follows immediately that $C_1(p_2,\ldots,p_r)$ must actually be a constant function. (This function is a complex analytic mapping from the compact complex analytic manifold M^{r-1} into the complex vector space $\mathbb{C}^{g\times g}$ of $g \times g$ complex matrices, so that by the proper mapping theorem its image is a compact complex analytic subvariety of $\mathbb{C}^{g\times g}$; but it follows directly from the maximum modulus theorem, and is proved in most of the standard texts on functions of several complex variables, that the only such subvarieties must consist of isolated points, from which the desired result follows immediately. Alternatively, any endomorphism $C\colon J(M) \longrightarrow J(M)$ must be determined by a $g \times g$ complex matrix C such that $C\cdot(I,\Omega) = (I,\Omega)\cdot D$ for some $2g \times 2g$ integer matrix D, as an immediate consequence of the condition that $C(\mathcal{L}) \subseteq \mathcal{L}$; thus the endomorphisms are necessarily a discrete subset of the vector space $\mathbb{C}^{g\times g}$ of all $g \times g$ matrices, so that even any continuous mapping from M^{r-1} into the space of all such endomorphisms must be a constant mapping.) The function $c_1(p_2,\ldots,p_r)$ is a complex analytic mapping $M^{r-1} \longrightarrow J(M)$, so that by induction

$$\psi(p_1,\ldots,p_r) = \sum_{j=1}^{r} C_j\varphi(p_j) + c$$

for some endomorphisms $C_j: J(M) \longrightarrow J(M)$ and some point $c \in J(M)$. If the mapping ψ merely depends on the divisor $p_1 + \ldots + p_r$, hence is invariant under permutations of the variables, it follows further that the endomorphisms C_j must all be equal, so that

$$\psi(p_1 + \ldots + p_r) = C\varphi(p_1 + \ldots + p_r) + c$$

for some endomorphism $C: J(M) \longrightarrow J(M)$ and some point $c \in J(M)$, where φ is the Jacobi homomorphism.

(b) That the image of a complex analytic mapping from a compact complex manifold into any analytic manifold is an analytic subvariety is a consequence of Remmert's proper mapping theorem, the proof of which can be found in most texts on functions of several complex variables. It should be emphasized that the image is an analytic subvariety, but not necessarily an analytic submanifold; an analytic subvariety is a closed subset which can be described locally as the set of common zeros of finitely many analytic functions.

(c) The subset $W^{\nu}_{r-1} + W_1 \subseteq W^{\nu}_r$ is an analytic subvariety of W^{ν}_r which will be called the <u>subvariety of gap points</u> of W^{ν}_r or the <u>gap subvariety</u> of W^{ν}_r ; its complement in W^{ν}_r is an open subset $\overset{\circ}{W}{}^{\nu}_r \subseteq W^{\nu}_r$ which will be called the <u>subset of nongap points</u> of W^{ν}_r . Of course all points of W^{ν}_r are gap points when $\nu = 1$, $r \geqq 1$, since $W_r = W_{r-1} + W_1$; so this concept is really only interesting for indices $\nu \geqq 2$. If all the points of W^{ν}_r are gap points then $\dim W^{\nu}_r = \dim W^{\nu}_{r-1} + 1$ provided that $1 \leqq \nu < r \leqq g-1$, a result that nicely complements Lemma 4 and indicates some of the usefulness of these concepts; but this is unfortunately not generally the case, so the situation is rather more complicated and not yet completely straightened out.

If a point $x \in J(M)$ is represented by a line bundle $\xi \in P(M)$, then $x \in W^{\nu}_r$ precisely when $\gamma(\xi\zeta^{r}_{p_0}) \geqq \nu$; and x is a gap point of W^{ν}_r precisely when $x - \varphi(p) \in W^{\nu}_{r-1}$ for some point $p \in M$, or equivalently, precisely when $\gamma(\xi\zeta^{-1}_{p}\zeta^{r}_{p_0}) \geqq \nu$ for some point $p \in M$. Consequently $x \in \overset{\circ}{W}{}^{\nu}_r$ if and only if the line bundle ξ representing x has the properties that $\gamma(\xi\zeta^{r}_{p_0}) = \nu$ and $\gamma(\xi\zeta^{-1}_{p}\zeta^{r}_{p_0}) = \nu-1$ for all points $p \in M$. If a nongap point $x \in \overset{\circ}{W}{}^{\nu}_r$ is represented as the image $x = \varphi(p_1 + \dots + p_r)$ of a divisor of degree r on M under the Jacobi homomorphism, then $\gamma(\zeta_{p_1}\cdots\zeta_{p_r}) = \nu$ and $\gamma(\zeta_{p_1}\cdots\zeta_{p_{i-1}}\zeta_{p_{i+1}}\cdots\zeta_{p_r}) = \nu-1$ for any index i , $1 \leq i \leq r$; hence for any such index i there must exist a meromorphic function

$f_i \in \Gamma(M, \mathcal{M})$ such that $\mathcal{D}(f_i) + p_1 + \dots + p_r \geqq 0$ but $\mathcal{D}(f_i) + p_1 + \dots + p_{i-1} + p_{i+1} + \dots + p_r \not\geqq 0$. It is then obvious that a general linear combination of these functions f_i will be a meromorphic function on M having as polar divisor precisely the divisor $-(p_1 + \dots + p_r)$. On the other hand if $x \in W_r^\nu$ is a gap point and $\nu \geqq 2$ then either $x \in W_r^{\nu+1}$ or x is represented by a line bundle $\xi \in P(M)$ such that $\gamma(\xi\zeta_{p_o}^r) = \nu$ and $\gamma(\xi\zeta_{p_1}^{-1}\zeta_{p_o}^r) = \nu$ for some point $p_1 \in M$; and in the latter case x can be represented as the image $x = \varphi(p_1 + \dots + p_r)$ of a divisor of degree r on M containing the point p_1, since $\nu \geqq 2$, and $\gamma(\zeta_{p_1} \cdots \zeta_{p_r}) = \gamma(\zeta_{p_2} \cdots \zeta_{p_r}) = \nu$. This last condition means that for any meromorphic function $f \in \Gamma(M, \mathcal{M})$ such that $\mathcal{D}(f) + p_1 + \dots + p_r \geqq 0$ necessarily $\mathcal{D}(f) + p_2 + \dots + p_r \geqq 0$ as well, and consequently there cannot exist any meromorphic function having as polar divisor precisely the divisor $-(p_1 + \dots + p_r)$. That explains the terminology, and indicates the extent to which this notion is a natural generalization of the classical notion of the Weierstrass gap sequence.

As an immediate consequence of the Riemann-Roch theorem, if a nongap point $x \in \overset{o}{W}{}_r^\nu$ is represented as the image $x = \varphi(p_1 + \dots + p_r)$ of a divisor of degree r on M, then $\gamma(\kappa\zeta_{p_1}^{-1} \cdots \zeta_{p_r}^{-1}) = \gamma(\kappa\zeta_{p_1}^{-1} \cdots \zeta_{p_{i-1}}^{-1}\zeta_{p_{i+1}}^{-1} \cdots \zeta_{p_r}^{-1}) = g-1-r+\nu$ for any index i; hence for any Abelian differential $\omega \in \Gamma(M, \mathcal{O}^{1,0})$ such that $\mathcal{D}(\omega) \geqq p_1 + \dots + p_{i-1} + p_{i+1} + \dots + p_r$ it necessarily follows that $\vartheta(\omega) \geqq p_1 + \ldots + p_r$ as well. Correspondingly if $x \in W_r^\nu$ is a gap point such that $x \notin W_r^{\nu+1}$ and if $\nu \geqq 2$, then x can be represented as the image $x = \varphi(p_1 + \ldots + p_r)$ of a divisor of degree r on M such that $\gamma(\kappa \zeta_{p_1}^{-1} \ldots \zeta_{p_r}^{-1}) = g-r+\nu-1$ and $\gamma(\kappa \zeta_{p_2}^{-1} \ldots \zeta_{p_r}^{-1}) = g-r+\nu$; consequently there must exist an Abelian differential $\omega \in \Gamma(M, \mathcal{O}^{1,0})$ such that $\vartheta(\omega) \geqq p_2 + \ldots + p_r$ but $\vartheta(\omega) \not\geqq p_1 + p_2 + \ldots + p_r$. Now the natural next step is to express these conditions in terms of the ranks of the matrices formed from the values of the Abelian differentials of a basis for $\Gamma(M, \mathcal{O}^{1,0})$ at the various points p_i , paralleling the classical form of the discussion of the Weierstrass gap sequence; but there are some difficulties in doing so, since the case in which all the points p_i are distinct and the cases in which there are some coincidences require separate treatment, so the continuation of this discussion will be postponed until these difficulties are resolved later in the course of these lectures.

It is also quite natural to consider the subvariety $W_{r+1}^{\nu+1} - W_1 \subseteq W_r^\nu$, and to seek to develop a corresponding notion of gap points in this context as well. However it is an immediate consequence of the Riemann-Roch theorem as expressed in formula (13) that $x \in W_{r+1}^{\nu+1} - W_1 \subseteq W_r^\nu$ precisely when $k-x$ is a gap point in the earlier sense of the subvariety W_s^μ , where $\mu = g-1- (r-\nu)$ and $s = 2g-2-r$ and the indices are such that $r, s \geqq 0$ and $\nu, \mu \geqq 1$; thus these two notions of gap points are really dual to one another, by appropriate use of the Riemann-Roch theorem, and it suffices to consider merely the one notion.

(d) On a hyperelliptic Riemann surface it follows from (25) that $W_r^\nu = W_{r-2\nu+2} - (\nu-1)\cdot e$ whenever $1 \leqq \nu \leqq r \leqq g-1$ and $r-2\nu+2 \geqq 0$; hence if in addition $r-2\nu+2 \geqq 1$ it follows that $W_r^\nu = W_1 + W_{r-2\nu+2} - (\nu-1)\cdot e = W_1 + W_{r-1}^\nu$, so that all the points of W_r^ν are gap points whenever $1 \leqq \nu \leqq r \leqq g-1$ and $r-2\nu+2 \geqq 1$. In the special case that $r-2\nu+2 = 0$, the subvariety W_r^ν consists merely of the point $-(\nu-1)\cdot e$ and $W_{r-1}^\nu = \emptyset$, so that the point $-(\nu-1)\cdot e$ is not a gap point of W_r^ν . Conversely if all the points of W_r^ν are gap points, so long as $W_{r-1}^\nu \neq \emptyset$, it is easy to see that the surface M is hyperelliptic; the details will be left to the reader. This indicates at any rate that there really are non-trivial nongap points on surfaces other than hyperelliptic surfaces.

References. The material discussed in this chapter has been taken primarily from the following sources.

[1] Hamilton, Richard S., Non-hyperelliptic Riemann surfaces, Jour. Differential Geom. 3(1969), 95-101.

[2] Martens, Henrik H., Torelli's theorem and a generalization for hyperelliptic surfaces, Comm. Pure Applied Math. 16 (1963), 97-110.

[3] -----, A new proof of Torelli's theorem, Annals of Math. 78(1963), 107-111.

[4] -----, On the varieties of special divisors on a Curve, I, Jour. reine Angew. Math. 227(1967), 111-120; II, Jour. reine Angew. Math. 233(1968), 89-100.

[5] -----, From the classical theory of Jacobian varieties, Proc. XV Scandinavian Congress of Mathematicians, Oslo, 1968. (Springer lecture notes 118, 74-98.

[6] -----, Three lectures in the classical theory of Jacobian varieties, Mimeographed notes, n.d.

[7] Matsusaka, T., On a theorem of Torelli, Amer. Jour. Math. 80(1958), 784-800.

[8] Mayer, A. L., Special divisors and the Jacobian variety, Math. Annalen 153(1964), 163-167.

[9] -----, On the Jacobi inversion theorem, (Ph.D. thesis, Princeton University, 1961).

[10] Weil, André, Zum Beweis des Torellischen Satzes, Nachrichten Akad. Wissenschaften Göttingen (1957), 33-53.

Other references, particularly to older literature, can be found in the bibliographies of these papers. The notation and properties of the subvarieties of positive divisors and of special positive divisors are given in several of these papers, in one form or another. The proof of Cliffords theorem used in these notes is that given in [4]; the extended form of this theorem, the assertion that if the subvariety W_r^ν attains the maximal dimension given in Theorem 7 for any indices in the range $2 \leqq \nu \leqq r \leqq g-2$ then the Riemann surface is hyperelliptic, can be found in [4] and in [1]. This will also be treated at a later point in these notes.

§3. Jacobi varieties and symmetric products of Riemann surfaces

(a) The restriction of the Jacobi homomorphism to the set of positive divisors of degree r can be viewed as a complex analytic mapping $\varphi: M^r \longrightarrow J(M)$; and it is evident that this mapping is really independent of the order of the factors in the Cartesian product M^r. This suggests introducing the <u>symmetric product</u> $M^{(r)}$, which is defined to be the quotient space $M^{(r)} = M^r/\mathfrak{S}_r$ of the compact complex analytic manifold M^r under the natural action of the symmetric group $\mathfrak{S}_r$ on r letters as the group of permutations of the factors in the Cartesian product M^r. That is to say, each permutation $\pi \in \mathfrak{S}_r$ of the integers $1,2,\ldots,r$ can be considered as defining a mapping $\pi: M^r \longrightarrow M^r$, by setting $\pi(p_1,\ldots,p_r) = (p_{\pi 1},\ldots,p_{\pi r})$; this exhibits $\mathfrak{S}_r$ as a group of analytic homeomorphisms of the manifold M^r, and the quotient space is by definition the symmetric product $M^{(r)}$. The points of $M^{(r)}$ can be considered as r-tuples of points of M, without regard to order, hence can be identified with positive divisors $\mathfrak{d} = p_1 + \ldots + p_r$ of degree r on M; consequently the symmetric product $M^{(r)}$ will also be called the <u>manifold of positive divisors of degree</u> r on the Riemann surface M.

<u>Theorem 9.</u> The r-fold symmetric product $M^{(r)}$ of a compact Riemann surface M has the structure of a compact complex analytic manifold of dimension r, such that the natural quotient mapping $\tau: M^r \longrightarrow M^{(r)}$ is a complex analytic mapping exhibiting the manifold M^r as an $(r!)$-sheeted branched analytic covering of the

manifold $M^{(r)}$.

Proof. It is evident that if none of the permutations in $\mathfrak{S}_r$ leave a point $(p_1,\ldots,p_r) \in M^r$ fixed, then these transformations will take a sufficiently small open neighborhood of that point into pairwise disjoint open subsets of M^r ; each of these sets can be viewed as a local coordinate neighborhood in the quotient space $M^{(r)} = M^r/\mathfrak{S}_r$, hence in that neighborhood $M^{(r)}$ will have the structure of an r-dimensional complex analytic manifold such that the natural quotient mapping $\tau\colon M^r \longrightarrow M^{(r)}$ is an (r!)-sheeted analytic covering mapping. Thus the only difficulty lies in the presence of fixed points of some of the transformations in $\mathfrak{S}_r$; these fixed points are of course the r-tuples $(p_1,\ldots,p_r) \in M^r$ such that not all of the points p_i are distinct. If there are s distinct points in this r-tuple the points can be renumbered so that coincidences occur in the form

$$(1)\qquad \begin{aligned} &p_1 = p_2 = \ldots = p_{\nu_1};\ p_{\nu_1+1} = p_{\nu_1+2} = \ldots = p_{\nu_1+\nu_2};\ \ldots ;\\ &\qquad p_{\nu_1+\ldots+\nu_{s-1}+1} = p_{\nu_1+\ldots+\nu_{s-1}+2} = \ldots = p_{\nu_1+\ldots+\nu_s} \end{aligned}$$

where $\nu_1 + \nu_2 + \ldots + \nu_s = r$. Select open coordinate neighborhoods U_i about the points p_i such that $U_i = U_j$ whenever $p_i = p_j$ but $U_i \cap U_j = \emptyset$ whenever $p_i \neq p_j$. The product $U = U_1 \times U_2 \times\ldots\times U_r$ is then an open coordinate neighborhood of the point $(p_1,\ldots,p_r)$ in M^r such that any permutation in $\mathfrak{S}_r$ either maps U onto itself or transforms U into an open subset of M^r disjoint from U ; indeed the permutations in $\mathfrak{S}_r$ mapping U onto itself form the subgroup $\mathfrak{S}_{\nu_1} \times \mathfrak{S}_{\nu_2} \times \dots \times \mathfrak{S}_{\nu_s} \subseteq \mathfrak{S}_r$ consisting of those permutations interchanging the first ν_1 indices among themselves, the second ν_2 indices among themselves, and so on. In order to conclude the proof it suffices to show that the quotient space $U/\mathfrak{S}_{\nu_1} \times \mathfrak{S}_{\nu_2} \times \dots \times \mathfrak{S}_{\nu_s}$ can be given the structure of a complex manifold of dimension r such that the natural mapping $U \longrightarrow U/\mathfrak{S}_{\nu_1} \times \mathfrak{S}_{\nu_2} \times \dots \times \mathfrak{S}_{\nu_s}$ is analytic. Letting z_i be local coordinates in the neighborhoods U_i , introduce the complex analytic mapping $\tau\colon U \longrightarrow \mathbf{C}^r$ defined by coordinate functions $\tau_i(z_1,\dots,z_r)$ of the form

$$
(2)\qquad
\begin{aligned}
\tau_1(z) &= z_1 + z_2 + \dots + z_{\nu_1}\\
\tau_2(z) &= z_1^2 + z_2^2 + \dots + z_{\nu_1}^2\\
\dots\quad & \quad\dots\\
\tau_{\nu_1}(z) &= z_1^{\nu_1} + z_2^{\nu_1} + \dots + z_{\nu_1}^{\nu_1}\\
\tau_{\nu_1+1}(z) &= z_{\nu_1+1} + z_{\nu_1+2} + \dots + z_{\nu_1+\nu_2}\\
\tau_{\nu_1+2}(z) &= z_{\nu_1+1}^2 + z_{\nu_1+2}^2 + \dots + z_{\nu_1+\nu_2}^2\\
\dots\quad & \quad\dots\\
\tau_r(z) &= z_{\nu_1+\dots+\nu_{s-1}+1}^{\nu_s} + \dots + z_{\nu_1+\dots+\nu_s}^{\nu_s} .
\end{aligned}
$$

The functions $\tau_1(z), \tau_2(z),\dots,\tau_{\nu_1}(z)$ are the elementary symmetric functions of the second kind in the variables $z_1, z_2, \ldots, z_{\nu_1}$; and as is well known, they determine an open complex analytic mapping $U_1 \times U_2 \times \ldots \times U_{\nu_1} \longrightarrow \mathbb{C}^{\nu_1}$ identifying the quotient space $U_1 \times U_2 \times \ldots \times U_{\nu_1} / \mathfrak{S}_{\nu_1}$ with an open subset of $\mathbb{C}^{\nu_1}$. The same observation can be made for each of the s blocks of coincident neighborhoods, and this evidently serves to conclude the proof of the theorem.

The particular local coordinates in the neighborhood of the point $p_1 + p_2 + \ldots + p_r$ in the manifold $M^{(r)}$ described by equation (2) will be used freely in the ensuing discussion. It should be noted that they depend on the choice of local coordinates in an open neighborhood of each of the distinct points p_i on the Riemann surface M.

(b) As already noted, the Jacobi homomorphism $\varphi: M^r \longrightarrow J(M)$ can be factored through the natural quotient mapping $\tau: M^r \longrightarrow M^{(r)}$, so that $\varphi = \psi \circ \tau$ for some complex analytic mapping $\psi: M^{(r)} \longrightarrow J(M)$; this mapping ψ will also be called the Jacobi mapping. In many ways the mapping ψ is more natural than the mapping φ, even though the manifold $M^{(r)}$ is more complicated than the manifold M^r; that will become more apparent as further properties of this mapping ψ are derived. For any index r in the range $1 \leq r \leq g-1$ the image of the mapping $\psi: M^{(r)} \longrightarrow J(M)$ is the proper analytic subvariety $W_r \subset J(M)$ of positive divisors of degree r, so that ψ can be viewed as an analytic mapping $\psi: M^{(r)} \longrightarrow W_r \subset J(M)$.

For any index r with $r \geqq g$ the image of the mapping $\psi: M^{(r)} \longrightarrow J(M)$ is the entire Jacobi variety; in particular the mapping $\psi: M^{(g)} \longrightarrow J(M)$ is an analytic mapping of the g-dimensional manifold $M^{(g)}$ onto the g-dimensional manifold $J(M)$, as a consequence of the Jacobi inversion theorem, and the problem of describing this analytic mapping in more detail can be viewed as an extended form of the Jacobi inversion problem. The analytic subvarieties W_r^ν of special positive divisors in W_r are transformed by the mapping ψ to analytic subvarieties $G_r^\nu = \psi^{-1}(W_r^\nu) \subseteq M^{(r)}$, which will also be called subvarieties of special positive divisors. Note that for a point $\mathfrak{d} \in M^{(r)}$ viewed as a divisor $\mathfrak{d} = p_1 + \dots + p_r$, the image $\psi(\mathfrak{d}) \in J(M)$ is also represented by the complex line bundle $\zeta_{p_1} \cdots \zeta_{p_r}\zeta_{p_0}^{-r}$ in the Picard variety; and this point $\psi(\mathfrak{d})$ lies in W_r^ν precisely when $\gamma(\zeta_{p_1} \cdots \zeta_{p_r}\zeta_{p_0}^{-r}\cdot\zeta_{p_0}^{r}) \geqq \nu$. Consequently the subvarieties $G_r^\nu \subseteq M^{(r)}$ can be defined by

$$G_r^\nu = \{p_1 + \dots + p_r \in M^{(r)} \mid \gamma(\zeta_{p_1} \cdots \zeta_{p_r}) \geqq \nu\} . \tag{3}$$

These subvarieties furnish a descending filtration

$$M^{(r)} = G_r^1 \supseteq G_r^2 \supseteq G_r^3 \supseteq \cdots$$

of the complex manifold $M^{(r)}$ by analytic subvarieties, which eventually terminate in the empty set.

Theorem 10(a). For any point $\vartheta = p_1 + \ldots + p_r \in M^{(r)}$ such that $\gamma(\zeta_{p_1} \cdots \zeta_{p_r}) = \nu$, the fibre $\psi^{-1}\psi(\vartheta)$ of the mapping $\psi: M^{(r)} \longrightarrow J(M)$ is an analytic subvariety of $M^{(r)}$ of dimension $\nu-1$ which can be represented as the image of a one-one analytic mapping of $\mathbb{P}^{\nu-1}$ into $M^{(r)}$.

Proof. That the fibre $\psi^{-1}\psi(\vartheta)$ is an analytic subvariety of $M^{(r)}$ is an immediate consequence of the analyticity of the mapping $\psi: M^{(r)} \longrightarrow J(M)$. It then suffices just to show that there exists a one-one analytic mapping $\rho: \mathbb{P}^{\nu-1} \longrightarrow M^{(r)}$ having as image precisely this analytic subvariety $\psi^{-1}\psi(\vartheta)$; the image $\rho(\mathbb{P}^{\nu-1}) = \psi^{-1}\psi(\vartheta)$ must then be an analytic subvariety of dimension $\nu-1$. Now it follows directly from Theorem 6 that the fibre $\psi^{-1}\psi(\vartheta)$ consists of the positive divisors of degree r linearly equivalent to the given divisor ϑ; and viewing these points as divisors on the Riemann surface M, they are just the divisors of arbitrary nontrivial holomorphic sections of the line bundle $\zeta_{\vartheta} = \zeta_{p_1} \cdots \zeta_{p_r}$. Letting $f_1, \ldots, f_\nu$ be a basis for the space of holomorphic sections of this line bundle, associate to any point $(c_1, \ldots, c_\nu) \in \mathbb{C}^\nu$ other than the origin the divisor

$$\rho(c_1, \ldots, c_\nu) = \vartheta(c_1 f_1 + \ldots + c_\nu f_\nu) \in M^{(r)};$$

there results a well defined mapping ρ from the complement of the origin in $\mathbb{C}^\nu$ into the manifold $M^{(r)}$, having as image precisely the fibre $\psi^{-1}\psi(\vartheta)$. Note that two nontrivial sections

$f', f'' \in \Gamma(M, \mathcal{O}(\zeta_{\mathfrak{d}}))$ have the same divisor if and only if they are constant multiples of one another, since the quotient f'/f'' is then a holomorphic function on all of M ; consequently $\rho(c'_1,\ldots,c'_\nu) = \rho(c''_1,\ldots,c''_\nu)$ if and only if $(c'_1,\ldots,c'_\nu) = (cc''_1,\ldots,cc''_\nu)$ for some nonzero complex constant c , so that ρ induces a one-one mapping $\rho\colon \mathbb{P}^{\nu-1} \longrightarrow M^{(r)}$ having as image precisely the fibre $\psi^{-1}\psi(\mathfrak{d})$.

It remains merely to show that this is a complex analytic mapping. Fixing a point in $\mathbb{P}^{\nu-1}$ represented by a vector $(c_1^o,\ldots,c_\nu^o) \in \mathbb{C}^\nu$, let $\rho(c_1^o,\ldots,c_\nu^o) = \mathfrak{d}(c_1^o f_1 + \ldots + c_\nu^o f_\nu) = p_1 + \ldots + p_r$ be a divisor in which coincidences among the points occur as in (1), so that this divisor can also be written as a divisor of distinct points in the form $\nu_1 p_1 + \nu_2 p_{\nu_1+1} + \ldots + \nu_s p_{\nu_1+\ldots+\nu_{s-1}+1}$; and choose disjoint coordinate neighborhoods centered at these distinct points in the Rieman surface M . In the coordinate neighborhood U_1 the section $c_1^o f_1 + \ldots + c_\nu^o f_\nu$ can be viewed as an ordinary complex analytic function, and its divisor in U_1 is just $\nu_1 \cdot p_1$; and if the constants $(c_1,\ldots,c_\nu)$ are sufficiently close to $(c_1^o,\ldots,c_\nu^o)$, it follows in a very familiar fashion that the section $c_1 f_1 + \ldots + c_\nu f_\nu$ will be an analytic function in U_1 having a divisor of total degree ν_1 there. Letting $z_1 + \ldots + z_{\nu_1}$ be the divisor of this section, where z_i denote the coordinates of these various points in terms of the chosen coordinate system in U_1 , it follows as usual from the Cauchy integral formula that

$$z_1^n + \dots + z_{\nu_1}^n = \frac{1}{2\pi i} \int_{\partial U_1} \frac{c_1 f_1'(z) + \dots + c_\nu f_\nu'(z)}{c_1 f_1(z) + \dots + c_\nu f_\nu(z)} z^n dz ;$$

here z is the local coordinate in U_1, the sections f_i are viewed as complex analytic functions $f_i(z)$ in U_1, and $f_i'(z)$ denotes the derivative of the function $f_i(z)$ with respect to the coordinate z. Now for $n = 1,2,\dots,\nu_1$ the expression $z_1^n + \dots + z_{\nu_1}^n$ is precisely one of the standard coordinates of the point $\vartheta(c_1 f_1 + \dots + c_\nu f_\nu)$ in $M^{(r)}$ as introduced in (2); and it is then obvious that the coordinates $t_1,\dots,t_{\nu_1}$ of the points $\rho(c_1,\dots,c_\nu) = \vartheta(c_1 f_1 + \dots + c_\nu f_\nu) \in M^{(r)}$ are complex analytic functions of the points $(c_1,\dots,c_\nu)$ in an open neighborhood of $(c_1^0,\dots,c_\nu^0)$. Considering the other coordinate neighborhoods $U_{\nu_1+1},\dots,U_{\nu_1+\dots+\nu_{s-1}+1}$ similarly, it finally follows that the mapping ρ is an analytic mapping in terms of the complex structure introduced on $M^{(r)}$, and that suffices to conclude the proof.

As an immediate consequence of this theorem, note that for $1 \leqq r \leqq g$ the analytic mapping $\psi: M^{(r)} \longrightarrow W_r$ is a one-one mapping from the complement of G_r^2 in $M^{(r)}$ onto the complement of W_r^2 in W_r; and further note that for $r \geq 2g-1$ the fibres of the analytic mapping $\psi: M^{(r)} \longrightarrow J(M)$ are analytic subvarieties of dimension $r-g$ which can be represented as images of one-one analytic mappings of $\mathbb{P}^{r-g}$ into the manifold $M^{(r)}$. Even more can easily be deduced as follows.

<u>Theorem 10(b).</u> At any point $\mathfrak{z} = p_1 + \ldots + p_r \in M^{(r)}$ such that $\gamma(\zeta_{p_1} \ldots \zeta_{p_r}) = \nu$, the differential of the analytic mapping $\psi: M^{(r)} \longrightarrow J(M)$ has rank given by

$$\text{rank } d\psi_{\mathfrak{z}} = r+1-\nu .$$

Proof. If coincidences in the divisor $\mathfrak{z} = p_1 + \ldots + p_r$ occur as in (1), that divisor can also be written as a divisor of distinct points in the form

$\mathfrak{z} = \nu_1 p_1 + \nu_2 p_{\nu_1+1} + \ldots + \nu_s p_{\nu_1+\ldots+\nu_{s-1}+1}$; and choosing disjoint coordinate neighborhoods centered at these distinct points on the Riemann surface M, introduce the standard coordinates on the symmetric product $M^{(r)}$ as in (2). The theorem will be proved simply by calculating the Jacobian matrix of the analytic mapping $\psi: M^{(r)} \longrightarrow J(M)$ at the point $\mathfrak{z}$, in terms of these coordinates on $M^{(r)}$ and the obvious coordinates on the Jacobi variety. Note that an Abelian integral of the first kind can be represented, uniquely up to an additive constant, by an analytic function $w_i(z)$ of the local coordinate in any of the coordinate neighborhoods chosen on the Riemann surface M ; and in terms of the local coordinate z centered at the point p_1, for example, this analytic function has a Taylor expansion

$$w_i(z) = w_i(p_1) + w_i'(p_1)z + \frac{1}{2!} w_i''(p_1)z^2 + \ldots .$$

Now if $z_1,\ldots,z_{\nu_1}$ are ν_1 points in this coordinate neighborhood, expressed in terms of the given local coordinate, then

$$w_i(z_1) +\ldots+ w_i(z_{\nu_1})$$

$$= \nu_1 w_i(p_1) + w_i'(p_1)(z_1 +\ldots+ z_{\nu_1}) + \frac{1}{2!} w_i''(p_1)(z_1^2 +\ldots+ z_{\nu_1}^2) + \ldots$$

$$= \nu_1 w_i(p_1) + w_i'(p_1)t_1 + \frac{1}{2!} w_i''(p_1)t_2 +\ldots+ \frac{1}{\nu_1!} w_i^{(\nu_1)}(p_1)\cdot t_{\nu_1} + \ldots ,$$

where $t_1,\ldots,t_{\nu_1}$ are the standard local coordinates in $M^{(r)}$ as in (2) and the remaining terms in the last series expansion involve higher powers of these local coordinates. The same construction can be carried out in the remaining coordinate neighborhoods as well. Thus for any divisor $z_1 + z_2 +\ldots+ z_r$ sufficiently near the given divisor $\mathcal{J} = p_1 + p_2 +\ldots+ p_r$ and represented by coordinates $(t_1,t_2,\ldots,t_r)$ in terms of the standard coordinate system chosen in a neighborhood of $\mathcal{J}$ on $M^{(r)}$, the analytic mapping $\psi\colon M^{(r)} \longrightarrow J(M)$ can be represented by the coordinate functions

$$\psi_i(t_1,\ldots,t_r) = w_i(z_1) + w_i(z_2) +\ldots+ w_i(z_r) =$$

$$= c_i + w_i'(p_1)t_1 + \frac{1}{2!} w_i''(p_1)t_2 +\ldots+ \frac{1}{\nu_1!} w_i^{(\nu_1)}(p_1)t_{\nu_1} +$$

$$+ w_i'(p_{\nu_1+1})\cdot t_{\nu_1+1} + \frac{1}{2!} w_i''(p_{\nu_1+1})t_{\nu_1+2} +\ldots+ \frac{1}{\nu_s!} w_i^{(\nu_s)}(p_{\nu_1+\ldots+\nu_{s-1}+1})\cdot t_r$$

+ higher order terms

for some constants c_i, where $1 \leq i \leq g$. The Jacobian matrix of the mapping ψ , as expressed in terms of these coordinates, is just the matrix of linear terms in this Taylor expansion; that matrix is just the $g \times r$ matrix having the following typical rows:

$$(4) \qquad d\psi_{\mathfrak{d}} = \Big(w_i'(p_1), \frac{1}{2!} w_i''(p_1), \ldots, \frac{1}{\nu_1!} w_i^{(\nu_1)}(p_1), \\ w_i'(p_{\nu_1+1}), \ldots, \frac{1}{\nu_s!} w_i^{(\nu_s)}(p_{\nu_1+\ldots\nu_{s-1}+1}) \Big)$$

$$\text{for } i = 1,\ldots,g .$$

Now for this matrix (4) observe that $g - \text{rank } d\psi_{\mathfrak{d}}$ is the dimension of the vector space consisting of those row vectors $(c_1,\ldots,c_g) \in \mathbb{C}^g$ such that $(c_1,\ldots,c_g)\cdot d\psi_{\mathfrak{d}} = 0$. For any such row vector, though, it is apparent from (4) that $\omega = c_1\omega_1 + \ldots + c_g\omega_g$ will be a holomorphic Abelian differential such that $\mathfrak{d}(\omega) \geq \mathfrak{d}$, where $\omega_i(z) = dw_i(z)$ are the basic Abelian differentials of the first kind. Therefore $g - \text{rank } d\psi_{\mathfrak{d}} = \gamma(\kappa\, \zeta_{\mathfrak{d}}^{-1})$; and since $\gamma(\kappa\, \zeta_{\mathfrak{d}}^{-1}) = \gamma(\zeta_{\mathfrak{d}}) - r+g-1$ as a consequence of the Riemann-Roch theorem, it follows that $\text{rank } d\psi_{\mathfrak{d}} = r+1 - \gamma(\zeta_{\mathfrak{d}})$, which was to be demonstrated.

Several almost immediate consequences of the combination of the two parts of this theorem deserve more detailed discussion.

Corollary 1 to Theorem 10. For any point $\mathfrak{d} = p_1 + \ldots + p_r \in M^{(r)}$ such that $\gamma(\zeta_{p_1} \cdots \zeta_{p_r}) = \nu$, the fibre $\psi^{-1}\psi(\mathfrak{d})$ of the mapping $\psi\colon M^{(r)} \longrightarrow J(M)$ is a complex analytic submanifold of $M^{(r)}$ which is analytically homeomorphic to $\mathbb{P}^{\nu-1}$.

Proof. As a consequence of Theorem 10(b) the Jacobian of the analytic mapping $\psi\colon M^{(r)} \longrightarrow J(M)$ has rank $r - (\nu-1)$ at each point of the fibre $\psi^{-1}\psi(\vartheta)$, hence the fibre must locally be contained in an analytic submanifold of $M^{(r)}$ of dimension $\nu-1$; but since as a consequence of Theorem 10(a) the fibre is an analytic subvariety of $M^{(r)}$ of dimension $\nu-1$, it must locally coincide with that submanifold, hence is itself an analytic submanifold of $M^{(r)}$. The one-one analytic mapping from $\mathbb{P}^{\nu-1}$ onto this submanifold $\psi^{-1}\psi(\vartheta)$, as in Theorem 10(a), is necessarily an analytic homeomorphism, and the desired result is thereby demonstrated.

<u>Corollary 2 to Theorem 10.</u> If $r \geqq 2g-1$ the analytic mapping $\psi\colon M^{(r)} \longrightarrow J(M)$ has the property that $\psi^{-1}(x)$ is an analytic submanifold of $M^{(r)}$ analytically homeomorphic to $\mathbb{P}^{r-g}$ for each point $x \in J(M)$.

Proof. If $r \geqq 2g-1$ then it follows from the Riemann-Roch theorem that $\gamma(\zeta_{p_1} \cdots \zeta_{p_r}) = r+1-g$ for any point $\vartheta = p_1 + \ldots + p_r \in M^{(r)}$; hence the desired result follows immediately from Corollary 1 to Theorem 10.

Actually somewhat more can be said in this case, as will be demonstrated later; the mapping $\psi\colon M^{(r)} \longrightarrow J(M)$ exhibits $M^{(r)}$ as a complex analytic fibre bundle over $J(M)$ with fibre $\mathbb{P}^{r-g}$.

Corollary 3 to Theorem 10. If $1 \leq r \leq g$ the analytic mapping $\psi: M^{(r)} \longrightarrow J(M)$ induces a complex analytic homeomorphism

$$\psi: M^{(r)} \backslash G_r^2 \xrightarrow{\cong} W_r \backslash W_r^2 .$$

(Here $A \backslash B$ denotes the set-theoretic difference between sets A and B, the complement of the subset B in A.)

Proof. For any divisor $\mathfrak{d} \in M^{(r)}$ such that $\mathfrak{d} \notin G_r^2$ necessarily $\gamma(\zeta_{\mathfrak{d}}) = 1$; hence as a consequence of Theorem 10(a) the restriction

$$\psi: M^{(r)} \backslash G_r^2 \longrightarrow W_r \backslash W_r^2$$

is a one-one analytic mapping between these two sets, and as a consequence of Theorem 10(b) the differential of this restriction has rank r at each point so is locally a complex analytic homeomorphism. That suffices to prove the desired result.

In the course of the proof of Theorem 10(b) the Jacobian matrix of the analytic mapping $\psi: M^{(r)} \longrightarrow J(M)$ was calculated quite explicitly, in terms of the standard local coordinates introduced on the manifolds $M^{(r)}$ and $J(M)$; the result of that calculation is useful by itself, and merits explicit mention. It is only natural to seek to express that result rather more intrinsically, though, at least avoiding the necessity of making particular choices of local coordinates on the manifolds $M^{(r)}$ and $J(M)$; and the interpretation of that result used in the proof of Theorem 10(b) suggests a convenient approach to such a reformulation.

In the representation of the Jacobi variety as a quotient group $J(M) = \mathbb{C}^g/\mathcal{L}$, the coordinates $(w_1,\ldots,w_g)$ in $\mathbb{C}^g$ provide convenient local coordinates at any point $x \in J(M)$; and in terms of these coordinates, a natural basis for the complex tangent space $T_x(J(M))$ of the manifold $J(M)$ at the point x is provided by the tangent vectors $\frac{\partial}{\partial w_1},\ldots,\frac{\partial}{\partial w_g}$, and dually a natural basis for the complex cotangent space $T_x^*(J(M))$ of the manifold $J(M)$ at the point x is provided by the covectors $dw_1,\ldots,dw_g$. The dual pairing $T_x(J(M)) \times T_x^*(J(M)) \longrightarrow \mathbb{C}$ is then given by

$$(a_1 \frac{\partial}{\partial w_1} +\ldots+ a_g \frac{\partial}{\partial w_g}) \times (b_1 dw_1 +\ldots+ b_g dw_g) = a_1 b_1 +\ldots+ a_g b_g ,$$

for arbitrary complex constants a_i, b_i . Now any covector $b_1 dw_1 +\ldots+ b_g dw_g \in T_x^*(J(M))$ extends to a unique group invariant covector field on the complex Lie group $J(M)$, namely to the holomorphic differential form $b_1 dw_1 +\ldots+ b_g dw_g \in \Gamma(J(M), \mathcal{O}^{1,0})$ on all of $J(M)$; and the restriction of this differential form to the analytic submanifold $W_1 \subseteq J(M)$ is a holomorphic differential form on W_1 , which under the Jacobi homeomorphism $\varphi: M \longrightarrow W_1$ induces the holomorphic differential form $b_1\omega_1 +\ldots+ b_g\omega_g \in \Gamma(M, \mathcal{O}^{1,0})$. There results a linear mapping $T_x^*(J(M)) \longrightarrow \Gamma(M, \mathcal{O}^{1,0})$, which is readily seen to be an isomorphism; hence the complex cotangent space $T_x^*(J(M))$ to the complex manifold $J(M)$ at any point x can naturally be identified with the space $\Gamma(M, \mathcal{O}^{1,0})$ of holomorphic Abelian differentials on M .

At any point $\vartheta \in M^{(r)}$ the analytic mapping $\psi: M^{(r)} \longrightarrow J(M)$ has a well defined differential $d\psi_{\vartheta}$, which is just the linear mapping

$$d\psi_{\vartheta} : T_{\vartheta}(M^{(r)}) \longrightarrow T_{\psi(\vartheta)}(J(M))$$

between the tangent spaces of these two manifolds induced by the mapping ψ. In terms of the natural bases provided in these tangent spaces by the standard local coordinates introduced on the manifolds $M^{(r)}$ and $J(M)$, the linear mapping $d\psi_{\vartheta}$ is that described by the Jacobian matrix (4); thus if $\vartheta = \nu_1 q_1 + \dots + \nu_s q_s \in M^{(r)}$ where $q_1, \dots, q_s$ are distinct points on the Riemann surface M, then the image of the mapping $d\psi_{\vartheta}$ is the linear subspace of $T_x(J(M))$ spanned by the vectors

$$w_1^{(m)}(q_n) \cdot \frac{\partial}{\partial w_1} + \dots + w_g^{(m)}(q_n) \cdot \frac{\partial}{\partial w_g}$$

for $1 \leq n \leq s$ and $1 \leq m \leq \nu_n$, where $w_i^{(m)}(q_n)$ are the derivatives of order m of the canonical Abelian differentials of the first kind on M, in terms of any local coordinates at the points $q_n \in M$, and $x = \psi(\vartheta)$. The linear subspace of $T_x^*(J(M))$ dual to the image of the mapping $d\psi_{\vartheta}$ is thus that consisting of those covectors $b_1 dw_1 + \dots + b_g dw_g$ such that

$$b_1 w_1^{(m)}(q_n) + \dots + b_g w_g^{(m)}(q_n) = 0$$

for all indices $1 \leq n \leq s$ and $1 \leq m \leq \nu_n$; and under the natural identification of $T_x^*(J(M))$ with $\Gamma(M, \mathcal{O}^{1,0})$, this corresponds to the linear subspace of $\Gamma(M, \mathcal{O}^{1,0})$ consisting of those Abelian differentials ω such that $\mathfrak{s}(\omega) \geqq \mathfrak{s}$. Thus (4) can be interpreted in the following form.

<u>Corollary 4 to Theorem 10.</u> For any point $\mathfrak{s} \in M^{(r)}$ the image of the differential of the mapping $\psi: M^{(r)} \longrightarrow J(M)$ is the linear subspace $d\psi_{\mathfrak{s}}(T_{\mathfrak{s}}(M^{(r)})) \subseteq T_{\psi(\mathfrak{s})}(J(M))$ dual to the subspace $L^*_{\mathfrak{s}} \subseteq T^*_{\psi(\mathfrak{s})}(J(M))$ defined by

$$(5) \qquad L^*_{\mathfrak{s}} = \{\omega \in \Gamma(M, \mathcal{O}^{1,0}) \mid \mathfrak{s}(\omega) \geqq \mathfrak{s}\}$$

with the natural identification $T^*_{\psi(\mathfrak{s})}(J(M)) = \Gamma(M, \mathcal{O}^{1,0})$ introduced above.

(c) It is an immediate consequence of Corollary 3 to Theorem 10 that the subvariety $W_r \subset J(M)$ is a regular analytic submanifold of $J(M)$ at any point not contained in W_r^2, whenever $1 \leq r \leq g-1$; and that leads to the problem of describing the singularities of the various subvarieties W_r^ν. As a matter of notation, the points of an analytic subvariety $V \subseteq J(M)$ at which V is a regular analytic submanifold of $J(M)$ are called the <u>regular points</u> of that subvariety, and the set of regular points of V will be denoted by $\mathcal{R}(V)$; the remaining points of V are called the <u>singular points</u> of that subvariety, and the set of singular points of V will be denoted by $\mathcal{S}(V)$. The singular locus $\mathcal{S}(V)$ is always a proper analytic subvariety of V. To any point $x \in V$ there can be associated the linear subspace $T^*_x(V) \subseteq T^*_x(J(M))$ spanned by all covectors of the form df_x , where f is any analytic function in an open neighborhood of the point x in $J(M)$ which vanishes identically on the subvariety V ; alternatively $T_x^*(V)$ can be described as the linear subspace of the cotangent space $T_x^*(J(M))$ formed by the differentials at x of all germs of analytic functions in the ideal of the analytic subvariety V at the point $x \in J(M)$. The natural dual to the subspace $T_x^*(V) \subseteq T_x^*(J(M))$ is a linear subspace $T_x(V) \subseteq T_x(J(M))$ which will be called the <u>tangent space</u> to the subvariety $V \subseteq J(M)$ at the point x ; and the dimension of the linear subspace $T_x(V)$ will be called the <u>imbedding dimension</u> of the variety V at the point x . The imbedding dimension of V at the point x can also be characterized as the smallest dimension of a local submanifold of $J(M)$ which contains the subvariety V in some open neighborhood of x in $J(M)$; the regular points of V are thus precisely those points at which the imbedding dimension of V is equal to the local dimension of V , the dimension of V in a small open neighborhood of x .

<u>Theorem 11.</u> (a) For any index r such that $1 \leq r \leq g-1$ the singular locus of the subvariety $W_r \subset J(M)$ is precisely the subvariety $W_r^2 \subset W_r$.

(b) For any indices $r \geq 1$, $\nu \geq 1$ such that W_r^ν is a proper analytic subvariety of $J(M)$, the subvariety $W_r^{\nu+1}$ is contained in the singular locus of W_r^ν ; indeed the analytic subvariety $W_r^\nu \subset J(M)$ has imbedding dimension equal to g at each point $x \in W_r^{\nu+1} \subset W_r^\nu$.

Proof. As already noted it follows from Corollary 3 to Theorem 10 that all points of W_r not contained in W_r^2 are regular points of W_r, hence that $\mathcal{S}(W_r) \subseteq W_r^2$; it is therefore sufficient merely to prove assertion (b), indeed merely to prove the last part of that assertion. Now recall from Lemma 3 that $W_r^{\nu+1} = W_r^{\nu} \ominus (W_1 - W_1)$; thus selecting a point $x \in W_r^{\nu+1}$, it follows that $x + W_1 - W_1 \subseteq W_r^{\nu}$, hence that $x + \varphi(p) - \varphi(q) \in W_r^{\nu}$ for any points $p, q \in M$. If f is any analytic function in an open neighborhood of x in $J(M)$ vanishing identically on W_r^{ν} in that neighborhood, and p is any point on the Riemann surface M, then $f(x + \varphi(p) - \varphi(q)) = 0$ identically as a function of q whenever q is sufficiently near p; and upon differentiating this identity with respect to q at the point $q = p$, in terms of a local coordinate system near p on the Riemann surface M, it follows that $d_x f \cdot \varphi'(p) = 0$ where $\varphi'(p)$ is the vector with components $\{w_i'(p)\}$ and the product is the ordinary vector inner product. Since the vectors $\varphi'(p)$ span the full vector space $\mathbb{C}^g$ as p varies over M, as a consequence of the familiar fact that the Abelian differentials $w_i'(z)dz = \omega_i(z)$ are linearly independent, it follows that $d_x f = 0$; but that must hold for all analytic functions f in an open neighborhood of x vanishing on W_r^{ν}, hence the imbedding dimension of the analytic subvariety W_r^{ν} at the point x is equal to g, and the proof is thereby concluded.

There remains the question whether the singular locus of W_r^ν is necessarily contained in $W_r^{\nu+1}$, when $\nu > 1$ and W_r^ν is a proper analytic subvariety of $J(M)$; at least some information can be obtained quite directly in the special case $\nu = 2$, $r \leqq g$. As a useful notational convention, the germ of an analytic subvariety $V \subseteq J(M)$ at a point $x \in J(M)$ will be denoted by $(V)_x$; thus $(V)_x$ indicates that the analytic subvariety V is only to be considered in arbitrarily small open neighborhoods of the point x in $J(M)$. Recall from Lemma 2 that $W_r^2 = W_{r-1} \ominus (-W_1) = \bigcap_{p \in M} [W_{r-1} + \varphi(p)]$; consequently $(W_r^2)_x = \bigcap_{p \in M} [(W_{r-1})_{x-\varphi(p)} + \varphi(p)]$, thus providing a representation of the germ of the analytic subvariety W_r^2 at any one of its points x as an intersection of appropriate translations of the germs of the analytic subvariety W_{r-1} at the various points $x - \varphi(p)$. Now note that $x - \varphi(p) \in W_{r-1}^2$ for all points $p \in M$ if and only if $x \in W_{r-1}^2 \ominus (-W_1) = W_r^3$; thus whenever $x \in W_r^2 \setminus W_r^3$ there will exist points $p \in M$ such that $x - \varphi(p) \in W_{r-1} \setminus W_{r-1}^2$, hence such that $x - \varphi(p)$ is a regular point of the subvariety W_{r-1}, since $\mathcal{R}(W_{r-1}) = W_{r-1} \setminus W_{r-1}^2$ as a result of Theorem 11 (a). For any such point p, the germ $(W_r^2)_x$ is contained in the germ $(W_{r-1})_{x-\varphi(p)} + \varphi(p)$, which is the germ of a complex analytic submanifold of dimension $r - 1$. The tangent space $T_x(W_r^2)$ is then contained in the tangent space of that submanifold, which is a linear space of dimension $r - 1$; and if $r \leqq g$ it follows that the imbedding dimension of W_r^2 at any point $x \in W_r^2 \setminus W_r^3$ is at most $r - 1$, hence is certainly less than g. Therefore the singularities of

W_r^2 which may be contained in $W_r^2 \backslash W_r^3$ are at least not so bad as the singularities contained in W_r^3 ; and in particular, if $r \leqq g$ W_r^3 <u>can be characterized intrinsically as the subvariety of</u> W_r^2 <u>consisting of those singular points at which the imbedding dimension of</u> W_r^2 <u>is equal to</u> g .

Of course this observation can readily be extended; for whenever $x \in W_r^2 \backslash W_r^3$, then $x - \varphi(p)$ will be a regular point of W_{r-1} for all but a finite number of points $p \in M$, hence the tangent space $T_x(W_r^2)$ is contained in the intersections of the tangent spaces of a number of germs of analytic submanifolds. At any regular point of the analytic subvariety W_{r-1} , the tangent space of W_{r-1} was determined explicitly in Corollary 4 to Theorem 10, so that something more can be said about the intersections of these tangent spaces also; but it is convenient to insert first a brief digression, to prepare the way for this calculation.

For any point $x \in W_r^2 \backslash W_r^3$, the fibre $\psi^{-1}(x)$ of the analytic mapping $\psi\colon M^{(r)} \longrightarrow J(M)$ is a complex analytic submanifold of $M^{(r)}$ analytically homeomorphic to $\mathbb{P}^1$, as a consequence of Corollary 1 to Theorem 10; thus the points of $\psi^{-1}(x)$ can be viewed as divisors $p_1(t) +\ldots+ p_r(t)$ depending analytically on a parameter $t \in \mathbb{P}^1$. It may very well happen that some of the points of these divisors remain fixed as t varies, so that these divisors can actually be written in the form $p_1(t) +\ldots+ p_s(t) + p_{s+1} +\ldots+ p_r$ for some index $s < r$. The divisors $p_1(t) +\ldots+ p_s(t)$ are obviously all linearly equivalent, and evidently

$\psi(p_1(t) + \dots + p_s(t)) = x' \in W_s^2 \setminus W_s^3$; thus $x = x' + x''$, where $x' \in W_s^2$ and $x'' = \psi(p_{s+1} + \dots + p_r) \in W_{r-s}$. If the index s is chosen to be as small as possible, this decomposition is obviously unique; and the condition that s be as small as possible is clearly that $x' \notin W_{s-1}^2 + W_1$. Using the terminology introduced in the notes to §2(c), the subset $W_{s-1}^2 + W_1 \subseteq W_s^2$ is the subvariety of <u>gap points</u> of W_s^2 , and its complement is the open subset $\mathring{W}_s^2$ of <u>nongap points</u> of W_s^2 . Thus any point $x \in W_r^2 \setminus W_r^3$ can be written uniquely in the form $x = x' + x''$, where $x' \in \mathring{W}_s^2$ and $x'' \in W_{r-s}$ for some index $s \leqq r$; and $x'' = \psi(p_{s+1} + \dots + p_r)$ for some uniquely determined divisor $p_{s+1} + \dots + p_r \in M^{(r-s)}$, so that $x'' \in W_{r-s} \setminus W_{r-s}^2$. In terms of this decomposition it is clear that any two divisors in the fibre $\psi^{-1}(x)$ of the analytic mapping $\psi: M^{(r)} \longrightarrow J(M)$ either coincide or have only the divisor $p_{s+1} + \dots + p_r$ in common. The points $p_{s+1}, \dots, p_r$ of this divisor are precisely those points $p \in M$ such that $x - \varphi(p) \in W_{r=1}^2 = \mathcal{S}(W_{r-1})$; so for any point $p \in M$ other than one of these, the translate $x - \varphi(p)$ is a regular point on the subvariety W_{r-1} .

<u>Lemma 5.</u> Consider a point $x \in W_r^2 \setminus W_r^3$, $1 \leq r \leq g$, which can be written in the form $x = x' + x''$ where $x' \in \mathring{W}_s^2$ and $x'' \in W_{r-s}$; and correspondingly, if $\xi \in H^1(M, \mathcal{O}^*)$ is the line bundle associated to any divisor in the fibre $\psi^{-1}(x)$ of the analytic mapping $\psi: M^{(r)} \longrightarrow J(M)$, write $\xi = \xi'\xi''$ where $c(\xi') = s$, $\gamma(\xi') = 2$, $c(\xi'') = r-s$, $\gamma(\xi'') = 1$. Then for any two points $p_1, p_2 \in M$ such that $x - \varphi(p_1)$ and $x - \varphi(p_2)$ are regular points of the subvariety

W_{r-1} , the tangent spaces $T_{x-\varphi(p_1)}(W_{r-1})$ and $T_{x-\varphi(p_2)}(W_{r-1})$ viewed as subspaces to the tangent space of $J(M)$ either coincide or intersect in a linear subspace of dimension $\gamma(\xi\xi') - 3+r-s$.

Proof. Since the tangent bundle to the Jacobi variety is trivial, the tangent spaces to $J(M)$ at various points can all be identified with one another canonically; that can be accomplished by translating all these tangent spaces to the same point of $J(M)$ using the group operation on $J(M)$. The dual cotangent spaces to $J(M)$ at various points can correspondingly be identified with one another canonically, and all can be identified with $\Gamma(M, \mathcal{O}^{1,0})$ as before. For each point $p_i \in M$ such that $x - \varphi(p_i)$ is a regular point of the subvariety W_{r-1} there will exist a unique positive divisor $\mathfrak{d}_i \in M^{(r-1)}$ such that $x = \varphi(p_i + \mathfrak{d}_i)$, and the tangent space to W_{r-1} at $x - \varphi(p_i) = \psi(\mathfrak{d}_i)$ can be identified with the image of the differential $d\psi_{\mathfrak{d}_i}$ of the analytic mapping $\psi: M^{(r-1)} \longrightarrow J(M)$ at the point $\mathfrak{d}_i \in M^{(r-1)}$; as a consequence of Corollary 4 to Theorem 10, the dual space to that tangent space is the linear subspace $L^*_{\mathfrak{d}_i} \subseteq T^*(J(M))$ defined by

$$L^*_{\mathfrak{d}_i} = \{\omega \in \Gamma(M, \mathcal{O}^{1,0}) \mid \mathfrak{d}(\omega) \geqq \mathfrak{d}_i\} .$$

Note that $\dim L^*_{\mathfrak{d}_i} = g-r+1$, and that as a consequence of the Riemann-Roch theorem $\dim L^*_{\mathfrak{d}_i+p_i} = g-r+1$ as well; so since $L^*_{\mathfrak{d}_i+p_i} \subseteq L^*_{\mathfrak{d}_i}$ these two spaces really coincide. Now the intersection of the tangent spaces to the subvariety W_{r-1} at the two regular points $x-\varphi(p_1)$ and $x-\varphi(p_2)$ is just the dual space to $L^*_{\vartheta_1} + L^*_{\vartheta_2} = L^*_{\vartheta_1+p_1} + L^*_{\vartheta_2+p_2}$; the dimension of that intersection is consequently $g-\dim(L^*_{\vartheta_1} + L^*_{\vartheta_2}) = g - \dim L^*_{\vartheta_1} - \dim L^*_{\vartheta_2} + {}$ $+ \dim(L^*_{\vartheta_1} \cap L^*_{\vartheta_2}) = 2(r-1) - g + \dim(L^*_{\vartheta_1} \cap L^*_{\vartheta_2})$. If the divisors $p_1 + \vartheta_1$ and $p_2 + \vartheta_2$ are distinct, their common terms are determined by the decomposition $x = x' + x''$ as discussed above; thus $p_1 + \vartheta_1 = \vartheta_1' + \vartheta''$ and $p_2 + \vartheta_2 = \vartheta_2' + \vartheta''$ where the divisors ϑ_1' and ϑ_2' have no common points and where $\zeta_{\vartheta_1'} = \zeta_{\vartheta_2'} = \xi'$ and $\zeta_{\vartheta''} = \xi''$. Then

$$
\begin{aligned}
L^*_{\vartheta_1} \cap L^*_{\vartheta_2} &= L^*_{\vartheta_1+p_1} \cap L^*_{\vartheta_2+p_2} \\
&= \{\omega \in \Gamma(M, \mathcal{O}^{1,0}) \mid \vartheta(\omega) \geqq \vartheta_1 + p_1 \text{ and } \vartheta(\omega) \geqq \vartheta_2 + p_2\} \\
&= \{\omega \in \Gamma(M, \mathcal{O}^{1,0}) \mid \vartheta(\omega) \geqq \vartheta_1' + \vartheta_2' + \vartheta''\} ,
\end{aligned}
$$

so that applying the Riemann-Roch theorem again

$$
\begin{aligned}
\dim(L^*_{\vartheta_1} \cap L^*_{\vartheta_2}) &= \gamma(\kappa(\xi')^{-2}(\xi'')^{-1}) \\
&= \gamma((\xi')^2\xi'') + g-1-r-s .
\end{aligned}
$$

Substituting this result into the preceding formula, it follows that the dimension of the intersection of the two tangent spaces is $\gamma((\xi')^2\xi'') - 3+r-s$ as desired, thus concluding the proof of the lemma.

If the two tangent spaces $T_{x-\varphi(p_1)}(W_{r-1})$ and $T_{x-\varphi(p_2)}(W_{r-1})$ in the preceding lemma intersect properly, that is to say, intersect in a linear subspace of dimension $2(r-1)-g$ in $\mathbb{C}^g$, then the two germs of manifolds $(W_{r-1})_{x-\varphi(p_1)}+\varphi(p_1)$ and $(W_{r-1})_{x-\varphi(p_2)}+\varphi(p_2)$ also intersect properly, in a complex analytic submanifold of $J(M)$ of dimension $2(r-1)-g$ containing the germ $(W_r^2)_x$. Since the dimension of any irreducible component of W_r^2 is not less than $2(r-1)-g$, as will shortly be demonstrated, then that intersection must coincide with W_r^2 in an open neighborhood of x, and consequently W_r^2 must be an analytic manifold at the point x. It follows from Lemma 5 that such a proper intersection occurs precisely when $\gamma(\xi\xi') = r+s+1-g$, with the notation as in the statement of the lemma; and since $c(\xi\xi') = r+s$, this is just the condition that the dimension $\gamma(\xi\xi') = \dim \Gamma(M, \mathcal{O}(\xi\xi'))$ have the least possible value. In other cases than this, the manifold germs $(W_{r-1})_{x-\varphi(p_1)} + \varphi(p_1)$ and $(W_{r-1})_{x-\varphi(p_2)} + \varphi(p_2)$ need not intersect properly, so their intersection need not be a submanifold of $J(M)$ near x; and the germ $(W_r^2)_x$ may be properly contained in the intersection. At any rate, the following does generally hold.

<u>Theorem 12(a).</u> For any point $x \in W_r^2 \setminus W_r^3$, $1 \le r \le g$, write $x = x' + x''$ where $x' \in \mathring{W}_s^2$ and $x'' \in W_{r-s}$; and let $\mathfrak{d} \in M^{(r)}$, $\mathfrak{d}' \in M^{(s)}$ be any positive divisors such that $\psi(\mathfrak{d}) = x$ and $\psi(\mathfrak{d}') = x'$. Then the imbedding dimension of the analytic variety W_r^2 at the point x is not greater than $\gamma(\zeta_{\mathfrak{d}}\zeta_{\mathfrak{d}'}) + r-s-3$.

Proof. Choosing any two points $p_1, p_2 \in M$ such that $x - \varphi(p_1)$ and $x - \varphi(p_2)$ are both regular points of the subvariety W_{r-1}, as already observed it follows that

$$T_x(W_r^2) \subseteq T_{x-\varphi(p_1)}(W_{r-1}) \cap T_{x-\varphi(p_1)}(W_{r-1}) ;$$

thus the imbedding dimension of W_r^2 at the point x, which is just $\dim T_x(W_r^2)$, is less than or equal to the dimension of the intersection of the two indicated tangent spaces to W_{r-1}. The desired result is then an immediate consequence of Lemma 5.

Recall from formula (17) of §2 that $\gamma(\zeta_\vartheta \zeta_{\vartheta'}) \leqq \frac{1}{2}(r+s)+1$ since $c(\zeta_\vartheta \zeta_{\vartheta'}) = r+s$; hence Theorem 12(a) yields the upper bound

$$\dim T_x(W_r^2) \leqq \frac{3}{2} r - \frac{1}{2} s - 2 . \tag{6}$$

This upper bound is best at nongap points $x \in \mathring{W}_r^2$, for which $r = s$ and hence $\dim T_x(W_r^2) \leqq r-2$; and of course since $r-2$ is the largest possible dimension for the subvariety W_r^2, and is actually attained for hyperelliptic Riemann surfaces, this is the best general estimate that could be expected. For more detailed results than provided by Theorem 12(a) in special cases, intersections of three or more tangent spaces should perhaps also be examined; that is somewhat involved, so in place of pursuing these detailed estimates let it suffice merely to state the obvious general consequence of these observations.

Theorem 12(b). For any point $x \in W_r^2 \backslash W_r^3$, $1 \leq r \leq g$, let L_x^* be the linear subspace of $\Gamma(M, \mathcal{O}^{1,0})$ spanned by all those holomorphic Abelian differentials ω such that $\vartheta(\omega) = \vartheta_1 + \vartheta_2$

where $\mathscr{D}_1 \in M^{(r)}$ and $\psi(\mathscr{D}_1) = x$. Then the imbedding dimension of the analytic variety W_r^2 at x is not greater than $g - \dim L_x^*$.

Proof. For any point $p \in M$ such that $x - \varphi(p)$ is a regular point of the subvariety W_{r-1}, the tangent space $T_x(W_r^2)$ is contained in the tangent space $T_{x-\varphi(p)}(W_{r-1})$, hence the dual space $T_x^*(W_r^2)$ contains the dual space $T_{x-\varphi(p)}^*(W_{r-1})$. These dual spaces are all viewed as subspaces of $T_x^*(J(M))$, and the latter space is as usual identified with the space $\Gamma(M, \mathcal{O}^{1,0})$; and if $x - \varphi(p) = \psi(\mathscr{D})$ for some divisor $\mathscr{D} \in M^{(r-1)}$, then

$$T_{x-\varphi(p)}^*(W_{r-1}) = L_{\mathscr{D}}^* = \{\omega \in \Gamma(M, \mathcal{O}^{1,0}) | \mathscr{D}(\omega) \geqq \mathscr{D}\}.$$

As noted in the proof of Lemma 5, this space necessarily coincides with the subspace

$$L_{\mathscr{D}+p}^* = \{\omega \in \Gamma(M, \mathcal{O}^{1,0}) | \mathscr{D}(\omega) \geqq \mathscr{D} + p\};$$

and the latter space consists of those Abelian differentials ω such that $\mathscr{D}(\omega) = \mathscr{D}_1 + \mathscr{D}_2$ where $\mathscr{D}_1 = \mathscr{D} + p \in M^{(r)}$ has the property that $\psi(\mathscr{D}_1) = x$. Conversely any differential $\mathscr{D}_1 \in M^{(r)}$ such that $\psi(\mathscr{D}_1) = x$ can be written in this form $\mathscr{D}_1 = \mathscr{D} + p$ for some $p \in M$ such that $x - \varphi(p) \in \mathcal{R}(W_{r-1})$. Then $T_x^*(W_r^2)$ must contain the linear span of all of these subspaces, and the proof is thereby concluded.

(d) For hyperelliptic Riemann surfaces it is quite easy to see that $W_r^{\nu+1}$ is precisely the singular locus of W_r^{ν} whenever $1 \leq \nu \leq r \leq g-1$; for recall from formula (25) of §2 that $W_r^{\nu} = W_{r-2\nu+2} - (\nu-1)e$ in this case, where $e \in J(M)$ is the hyperelliptic point, hence applying Theorem 11(a) note that $\mathcal{S}(W_r^{\nu}) = \mathcal{S}(W_{r-2\nu+2}) - (\nu-1)e = W^2_{r-2\nu+2} - (\nu-1)e = (W_{r-2\nu} - e) - (\nu-1)e = W_{r-2\nu} - \nu e = W_r^{\nu+1}$. In particular, for a hyperelliptic Riemann surface the subvarieties W_r^{ν} of special positive divisors are fully determined by the structure of the singular loci of the subvarieties W_r of positive divisors.

Using the results obtained in part (c) of this section, it is easy to derive the extension of Clifford's theorem that has already been mentioned, the result that hyperelliptic Riemann surfaces are characterized by the subvarieties W_r^{ν} being of the maximum dimension described in Theorem 7.

Theorem 13. If $\dim W_r^{\nu} = r-2\nu+2$ for some pair of indices ν,r in the range $2 \leqq \nu \leqq \frac{r}{2}+1$, $r \leqq g-2$, for a Riemann surface of genus g , then that surface is hyperelliptic.

Proof. If $\dim W_r^{\nu} = r-2\nu+2$ for some index $\nu > 2$, then since $\dim W_r^{\nu} < \dim W_{r-1}^{\nu-1}$ by Lemma 4 and $\dim W_{r-1}^{\nu-1} \leqq r-2\nu+3$ by Theorem 7, it follows that $\dim W_{r-1}^{\nu-1} = r-2\nu+3$; this is the same hypothesis as that of the present theorem but for indices $\nu-1, r-1$ in place of ν,r , and by repeating the argument as necessary it evidently suffices to prove the theorem for the special case $\nu = 2$

only. Assume therefore that $\dim W_r^2 = r-2$ for some index r in the range $2 \leqq r \leqq g-2$. If all the points of W_r^2 are gap points, then $W_r^2 = W_{r-1}^2 + W_1$ and hence $\dim W_{r-1}^2 \geqq \dim W_r^2 - 1 = r-3$; and since $\dim W_{r-1}^2 \leqq r-3$ by Theorem 7 it follows that $\dim W_{r-1}^2 = r-3$, which is again the same hypothesis but for the index $r-1$ in place of r. If all the points of W_{r-1}^2 are gap points, the argument can be repeated again; hence eventually either $\dim W_2^2 = 0$, hence $W_2^2 \neq \emptyset$ and the surface is hyperelliptic as desired, or $\mathring{W}_r^2 \neq \emptyset$ for some index r. In the latter case, choosing a regular point $x \in \mathring{W}_r^2$ at which the analytic subvariety W_r^2 has dimension $r-2$ and a divisor $\mathfrak{d} \in M^{(r)}$ such that $\psi(\mathfrak{d}) = x$, it follows from Theorem 12(a) that $r-2 = \dim W_r^2 = \dim T_x(W_r^2) \leqq \gamma(\zeta_{\mathfrak{d}}^2) - 3$, and hence $\gamma(\zeta_{\mathfrak{d}}^2) \geqq r+1$; but then $\zeta_{\mathfrak{d}}^2 \in W_{2r}^{r+1}$ so that $W_{2r}^{r+1} \neq \emptyset$, and it follows from Theorem 8 (Clifford's Theorem) that the surface is hyperelliptic in this case as well. That suffices to conclude the proof.

(e) Returning again to the consideration of the Jacobi mapping $\psi: M^{(r)} \longrightarrow J(M)$, note that the differential of this mapping can be viewed as a complex analytic mapping from the tangent bundle of the manifold $M^{(r)}$ to the restriction of the tangent bundle of the manifold $J(M)$ to the image subvariety $W_r = \psi(M^{(r)}) \subseteq J(M)$. Recall from Theorem 10(b) that at any point $\mathfrak{d} \in M^{(r)}$

$$\operatorname{rank} d\psi_{\mathfrak{d}} = r+1 - \gamma(\zeta_{\mathfrak{d}}) ;$$

hence the analytic subvarieties $G_r^\nu \subseteq M^{(r)}$ can be described in terms of the analytic bundle mapping $d\psi$ as

$$(7) \qquad G_r^\nu = \{ \mathfrak{s} \in M^{(r)} \,|\, \text{rank } d\psi_{\mathfrak{s}} \leqq r+1-\nu \} .$$

This is actually a very convenient description of the analytic subvarieties G_r^ν, both locally and globally. To examine this description locally at first, choosing local coordinates $(t_1,\ldots,t_r)$ in an open neighborhood of a point $\mathfrak{s} \in M^{(r)}$ and local coordinates $(w_1,\ldots,w_g)$ in an open neighborhood of the image $\psi(\mathfrak{s}) \in J(M)$, the mapping ψ is represented by coordinate functions $w_i(t_1,\ldots,t_r)$, $1 \leq i \leq g$, and its differential by the $g \times r$ matrix

$$d\psi(t) = \left\{ \frac{\partial w_i}{\partial t_j} \right\} ,$$

the entries of which are analytic functions in the given coordinate neighborhood of $M^{(r)}$; in that neighborhood, G_r^ν is the analytic subvariety consisting of those points at which this matrix of analytic functions has rank at most $r+1-\nu$. Thus the subvariety G_r^ν can be described locally quite explicitly as the set of common zeros of all subdeterminants of rank $r-\nu+2$ that can be formed from the matrix $d\psi$. This description is rather redundant, though, and can be improved by making use of the following observation.

<u>Lemma 6.</u> Viewing the set $\mathbf{C}^{g\times r}$ of $g\times r$ complex matrices Z as the space of gr complex variables, in an open neighborhood of any matrix $A \in \mathbf{C}^{g\times r}$ of rank n the subset of those matrices

$Z \in \mathbf{C}^{g\times r}$ such that $\operatorname{rank} Z \leqq n$ is a complex analytic submanifold of $\mathbf{C}^{g\times r}$ of dimension $gr - (g-n)(r-n)$.

Proof. By suitably renumbering the rows and columns of the matrices it can be assumed that the principal $n \times n$ minor of the matrix A is nonsingular; the principal $n \times n$ minors of all matrices Z sufficiently near A will then also be nonsingular. For any indices i,j in the range $n+1 \leqq i \leqq g$, $n+1 \leqq j \leqq r$, let $f_{ij}(Z)$ denote the determinant of the $(n+1) \times (n+1)$ submatrix of Z formed by adjoining the appropriate elements in row i and column j of Z to the principal $n \times n$ minor; there are thus $(g-n)(r-n)$ of these functions. Note that the variable z_{ij} (the coordinate in row i and column j of the matrix Z) appears only in the function $f_{ij}(Z)$, and that $\partial f_{ij}(Z)/\partial z_{ij}$ is the determinant of the principal $n \times n$ minor of the matrix Z and hence is nonzero for all matrices Z sufficiently near A; therefore the functions $f_{ij}(Z)$ can be used in place of the coordinates z_{ij} as part of a regular coordinate system in $\mathbf{C}^{g\times r}$ near A, and consequently the set of common zeros of the functions $f_{ij}(Z)$ form a complex analytic submanifold of $\mathbf{C}^{g\times r}$ of dimension $gr - (g-n)(r-n)$ near A. It is obvious that this submanifold is contained in the subvariety consisting of those matrices Z such that $\operatorname{rank} Z \leqq n$; to conclude the proof it is only necessary to show that conversely any matrix Z near A belonging to this submanifold must have the property that $\operatorname{rank} Z \leqq n$. Considering the $g \times (n+1)$ matrix consisting of columns

$1,2,\dots,n,j$ of the matrix Z, the equations $f_{ij}(Z) = 0$, $i = n+1,\dots,g$, show that rows $1,2,\dots,n,i$ are linearly dependent; and since the first n rows are linearly independent, each succeeding row must be a linear combination of the first n rows. The entire $g\times(n+1)$ matrix must therefore have rank n; and hence the columns are linearly dependent. Again the first n columns of Z are linearly independent, hence each succeeding column must be a linear combination of the first n columns, so that $\operatorname{rank} Z \leqq n$; and that suffices to conclude the proof.

Theorem 14 (a). If V is an irreducible component of the analytic subvariety $G_r^\nu \subseteq M^{(r)}$ and if V is not entirely contained within $G_r^{\nu+1}$ then

$$\dim V \geqq r\nu - (\nu-1)(g+\nu-1) .$$

Proof. Select a point $\mathfrak{d} \in V \subseteq G_r^\nu$ such that $\mathfrak{d}$ is not contained in any other irreducible component of G_r^ν and $\mathfrak{d} \notin G_r^{\nu+1}$; and in terms of any local coordinates $(t_1,\dots,t_t)$ near $\mathfrak{d}$ in $M^{(r)}$ and $(w_1,\dots,w_g)$ near $\psi(\mathfrak{d})$ in $J(M)$, write the differential of the Jacobi mapping $\psi\colon M^{(r)} \longrightarrow J(M)$ as the matrix of analytic functions $d\psi = \{\partial w_i/\partial t_j\}$. Near the point $\mathfrak{d}$ the subvariety V consists of those points at which $\operatorname{rank} d\psi \leqq r+1-\nu$, as a consequence of (7); and at the point $\mathfrak{d}$ itself $\operatorname{rank} d\psi_{\mathfrak{d}} = r+1-\nu$ since $\mathfrak{d} \notin G_r^{\nu+1}$. Now it follows from Lemma 6 that the set of matrices near $d\psi_{\mathfrak{d}}$ having $\operatorname{rank} \leqq r+1-\nu$ is a complex analytic submanifold of the matrix space $\mathbb{C}^{g\times r}$ which can be defined by the vanishing of

$(g-r-1+\nu)(r-r-1+\nu)$ local coordinate functions; hence the subvariety V near $\mathfrak{s}$ can be defined by the vanishing of $(g-r-1+\nu)(\nu-1)$ analytic functions, from which it follows as the result of a familiar theorem in complex analysis that

$$\dim V \geqq r - (g-r-1+\nu)(\nu-1) = r\nu - (\nu-1)(g+\nu-1) .$$

That serves to conclude the proof of the theorem.

<u>Corollary 1 to Theorem 14.</u> For any nonempty irreducible component V of the analytic subvariety $G_r^\nu \subseteq M^{(r)}$,

$$\dim V \geqq r\nu - (\nu-1)(g+\nu-1)$$

whenever $1 \leqq \nu \leqq r \leqq g-1$.

Proof. In view of Theorem 14 it suffices to show that if V is an irreducible component of G_r^ν with $r \leqq g-1$ then $V \not\subseteq G_r^{\nu+1}$. Suppose conversely that $V \subseteq G_r^{\nu+1}$. Then $\psi(V)$ is an irreducible component of the analytic subvariety $\psi(G_r^\nu) = W_r^\nu \subset J(M)$ such that $\psi(V) \subseteq W_r^{\nu+1}$. Since $W_r^{\nu+1} = W_r^\nu \ominus (W_1-W_1)$ by Lemma 3, it follows that

$$\psi(V) \subseteq \psi(V) + W_1-W_1 \subseteq W_r^{\nu+1} + W_1-W_1 \subseteq W_r^\nu ;$$

and since $\psi(V) + W_1-W_1$ is also an irreducible subvariety of W_r^ν , necessarily $\psi(V) = \psi(V) + W_1-W_1$. This in turn can only happen when $\psi(V) = J(M)$, as noted in the proof of Lemma 4; and that is impossible, since $W_r^\nu \subset J(M)$ when $r \leqq g-1$. This contradiction serves to complete the proof of the corollary.

It should also be noted that while this theorem provides a lower bound for the dimension of any irreducible component of the analytic subvariety $G_r^\nu \subseteq M^{(r)}$, it does not assert that the subvariety G_r^ν is necessarily nonempty when that lower bound is nonnegative. However this extension of the theorem does hold, and can be derived as a global consequence of the description of the subvarieties G_r^ν provided by equation (7); but this really involves more the topological than the analytical consequences of this form of the description of the subvarieties G_r^ν, so that while the idea of the proof can be given very simply here, the details must be postponed until after the relevant topological properties have been discussed. In terms of any local coordinates $(t_1,\ldots,t_r)$ in a coordinate neighborhood on the manifold $M^{(r)}$ and some canonical local coordinates $(w_1,\ldots,w_g)$ on the Jacobi variety $J(M)$, the Jacobi mapping $\psi: M^{(r)} \longrightarrow J(M)$ is represented by coordinate functions $w_i(t_1,\ldots,t_r)$, $1 \le i \le g$, as noted above. Now the canonical local coordinates on the Jacobi variety are any local coordinates derived from the natural coordinates on $\mathbb{C}^g$ when the Jacobi variety is represented as the quotient group $J(M) = \mathbb{C}^g/\mathcal{L}$; so the choice of another set of canonical local coordinates on $J(M)$ merely alters the coordinate functions representing the Jacobi mapping by the addition of constant terms, corresponding to appropriate points of the lattice subgroup $\mathcal{L} \subset \mathbb{C}^g$. The differential forms $dw_i = \sum_j \frac{\partial w_i}{\partial t_j} dt_j$ are therefore well defined and independent of the choice of canonical local coordinates on $J(M)$. Thus the g rows of the matrix $d\psi = \{\partial w_i/\partial t_j\}$ can be viewed as g holomorphic differential forms (or holomorphic cotangent vector fields) over the manifold $M^{(r)}$. Now recalling equation (7), if the subvariety G_r^ν is empty these g cotangent vector fields will span a subspace of the cotangent space of $M^{(r)}$ of dimension $\geqq r+2-\nu$ at each point of $M^{(r)}$; in particular, if G_r^2 is empty these g cotangent vector fields will span the full cotangent space of the manifold $M^{(r)}$ at each point. Thus if there is a topological obstruction to the existence of g continuous covector fields over the manifold $M^{(r)}$ which span the entire cotangent space of $M^{(r)}$ at each point, then necessarily G_r^2 is nonempty; or more generally, if there are topological obstructions to the existence of g continuous vector fields over the manifold $M^{(r)}$ which span a subspace of the cotangent space of dimension $\geqq r+2-\nu$ at each point of $M^{(r)}$, then necessarily G_r^ν is nonempty.

(f) Recall from Theorem 10 that the fibres of the Jacobi mapping $\psi\colon M^{(r)} \longrightarrow J(M)$ are complex analytic submanifolds of $M^{(r)}$ analytically homeomorphic to complex projective spaces of various dimensions. To round out the description of this mapping, something further should be added to indicate the extent to which the Jacobi mapping exhibits $M^{(r)}$ as a local product of $J(M)$ and the appropriate projective space.

Theorem 15. For any sufficiently small relatively open subset V of the analytic subvariety $W_r^\nu \backslash W_r^{\nu+1} \subseteq J(M)$ there exists a one-one analytic mapping

$$\lambda_V\colon V \times \mathbb{P}^{\nu-1} \longrightarrow \psi^{-1}(V) \subseteq M^{(r)}$$

such that $\psi\circ\lambda_V\colon V \times \mathbb{P}^{\nu-1} \longrightarrow V$ is the natural projection.

Proof. The proof of the desired result is a rather straightforward modification of the proof of Theorem 10(a). First choose disjoint open subsets U_i of the Riemann surface M such that the restricted Jacobi mapping $\varphi\colon U_1 \times \dots \times U_g \longrightarrow U \subseteq J(M)$ is a complex analytic homeomorphism; that there exist such sets is an immediate consequence of the observation that $\varphi = \psi\tau$ where $\tau\colon M^g \longrightarrow M^{(g)}$ is a branched analytic covering and $\psi\colon M^{(g)}\backslash G_g^2 \longrightarrow J(M)\backslash W_g^2$ is a complex analytic homeomorphism, G_g^2 and W_g^2 being proper analytic subvarieties. Fixing points $p_k^o \in U_k$ and introducing the divisor $\mathfrak{d} = p_1^o + \dots + p_g^o \in M^{(g)}$, it follows that for any fixed line bundle $\eta \in P(M)$ the mapping $(p_1,\dots,p_k) \longrightarrow \eta\zeta_{\mathfrak{d}}\, \zeta_{p_1}^{-1} \cdots \zeta_{p_g}^{-1}$ is a complex analytic homeomorphism from $U_1 \times \dots \times U_g$ onto an open neighborhood of the point η in the manifold $P(M)$; this provides a very convenient system of local coordinates in a neighborhood of any fixed point of $P(M)$, or equivalently of course, in a neighborhood of any fixed point of $J(M)$. In terms of this parametrization, the fibre of the Jacobi mapping $\psi\colon M^{(r)} \longrightarrow J(M)$ over the point of $J(M)$ described by the parameters $(p_1,\dots,p_k)$ consists precisely of the divisors of all the holomorphic sections $f \in \Gamma(M, \mathcal{O}(\eta\zeta_{\vartheta}\, \zeta_{p_1}^{-1} \cdots \zeta_{p_g}^{-1}\zeta_{p_o}^{r}))$. Recalling the canonical isomorphism

$$\Gamma(M, \mathcal{O}(\eta\zeta_{\vartheta}\, \zeta_{p_1}^{-1}\cdots\zeta_{p_g}^{-1}\zeta_{p_o}^{r})) \cong \{f \in \Gamma(M, \mathcal{O}(\eta\zeta_{\vartheta}\, \zeta_{p_o}^{r})) \mid \vartheta(f) \geqq p_1+\cdots+p_g\} ,$$

then upon choosing a basis $f_1,\ldots,f_n$ for the vector space $\Gamma(M, \mathcal{O}(\eta\zeta_{\vartheta}\, \zeta_{p_o}^{r}))$ it follows that the fibre of the Jacobi mapping $\psi\colon M^{(r)} \longrightarrow J(M)$ over the point $(p_1,\ldots,p_r)$ can be described equivalently as the set of divisors of the form

$$\vartheta\,(c_1f_1 +\cdots+ c_nf_n) - (p_1 +\cdots+ p_g) \in M^{(r)}$$

where $c_1,\ldots,c_n$ are any complex constants such that $\sum_{j=1}^{n} c_jf_k(p_k) = 0$ for $k = 1,\ldots,g$. Note that the sections f_j are here viewed as ordinary complex analytic functions in each coordinate neighborhood U_k.

Now if $\eta \in W_r^{\nu}\backslash W_r^{\nu+1}$ so that $\gamma(\eta\zeta_{p_o}^{r}) = \nu$, it follows that the $n\times g$ matrix $\{f_j(p_k^o)\}$ has rank $n-\nu$; and near η the subset $W_r^{\nu}\backslash W_r^{\nu+1}$ can be described as the analytic subvariety $V \subseteq U_1\times\cdots\times U_g$ consisting of those points $(p_1,\ldots,p_g)$ such that $\operatorname{rank}\{f_j(p_k)\} \leqq n-\nu$. It can be assumed, renumbering the sections f_j and shrinking the neighborhoods U_k of p_k^o, if necessary, that the $(n-\nu)\times g$ submatrix $\{f_j(p_k)\}$, $1 \leq j \leq n-\nu$, $1 \leq k \leq n-\nu$, is nonsingular for $p_k = p_k^o$ and hence for all $p_k \in U_k$, $1 \leq k \leq n-\nu$. The linear equations $\sum_{j=1}^{n} c_jf_j(p_k) = 0$, $1 \leq k \leq n-\nu$, then have unique solutions $c_1,\ldots,c_{n-\nu}$ for arbitrary constant values $c_{n-\nu+1},\ldots,c_n$; and Cramer's formula shows that these solutions $c_1,\ldots,c_{n-\nu}$ are complex analytic functions of the points $(p_1,\ldots,p_g) \in U_1 \times\ldots\times U_g$. It should be remarked again that the sections f_j are viewed as ordinary complex analytic functions in each coordinate neighborhood U_k, by choosing any fixed trivialization of the line bundle $\eta\zeta_{\vartheta}\zeta_{p_0}^r$ over each neighborhood. Taking any ν linearly independent constant vectors $(c^i_{n-\nu+1},\ldots,c^i_n)$, $1 \leq i \leq \nu$, there thus exist ν vectors of holomorphic functions $(c^i_1(p_1,\ldots,p_g),\ldots,c^i_n(p_1,\ldots,p_g))$, $i \leq i \leq \nu$, in $U_1 \times\ldots\times U_g$ such that for any fixed point $(p_1,\ldots,p_g)$ the sections $\sum_{j=1}^{n} c^i_j(p_1,\ldots,p_g)f_j \in \Gamma(M,\mathcal{O}(\eta\zeta_{\vartheta}\zeta_{p_0}^r))$, $1 \leq i \leq \nu$, are linearly independent and $\sum_{j=1}^{n} c^i_j(p_1,\ldots,p_g)f_j(p_k) = 0$ for $1 \leq i \leq \nu$, $1 \leq k \leq n-\nu$; and moreover whenever $(p_1,\ldots,p_g) \in V$ then also $\sum_{j=1}^{n} c^i_j(p_1,\ldots,p_g)f_j(p_k) = 0$ for $1 \leq i \leq \nu$, $n-\nu+1 \leq k \leq g$. Introducing then the mapping

$$\lambda_V \colon V \times \mathbf{C}^{\nu} \longrightarrow M^{(r)}$$

defined by

$$\lambda_V(p_1,\ldots,p_g;t_1,\ldots,t_\nu) = \vartheta\Big(\sum_{i=1}^{\nu}\sum_{j=1}^{n} t_i c^i_j(p_1,\ldots,p_g)f_j\Big) - (p_1+\ldots+p_g)$$

it is evident from the proof of Theorem 10(a) and from the construction that this is a complex analytic mapping which induces a one-one complex analytic mapping from the analytic variety $V \times \mathbb{P}^{\nu-1}$ onto the subset $\psi^{-1}(V) \subseteq M^{(r)}$. That suffices to conclude the proof of the theorem.

If the subvariety $\psi^{-1}(V) \subseteq M^{(r)}$ has rather bad singularities, a one-one analytic mapping from another analytic variety onto $\psi^{-1}(V)$ need not have an analytic inverse, so need not be an analytic homeomorphism; but ignoring these complications, the following simple consequences of this theorem easily arise.

Corollary 1 to Theorem 15. For any sufficiently small relatively open subset $V \subseteq W_r^{\nu} \setminus W_r^{\nu+1}$ such that $\psi^{-1}(V) \subseteq \mathcal{R}(G_r^{\nu})$, the set $\psi^{-1}(V)$ is analytically homeomorphic to the product $V \times \mathbb{P}^{\nu-1}$.

Proof. If $\psi^{-1}(V) \subseteq \mathcal{R}(G_r^{\nu})$ so that the set $\psi^{-1}(V)$ is itself a complex manifold, the one-one analytic mapping $\lambda_V: V \times \mathbb{P}^{\nu-1} \longrightarrow \psi^{-1}(V)$ is an analytic homeomorphism, hence the corollary follows trivially.

Corollary 2 to Theorem 15. If $r \geqq 2g-1$ the analytic mapping $\psi: M^{(r)} \longrightarrow J(M)$ exhibits the manifold $M^{(r)}$ as a locally trivial analytic fibration over the manifold $J(M)$ with fibre $\mathbb{P}^{r-g}$.

Proof. If $r \geqq 2g-1$ then $W_r^{r-g+1} = J(M)$ and $W_r^{r-g+2} = \emptyset$; it then follows from Corollary 1 that for any sufficiently small open subset $V \subseteq J(M)$ the inverse image $\psi^{-1}(V) \subseteq M^{(r)}$ is analytically homeomorphic to the product $V \times \mathbb{P}^{r-g}$. This is precisely the desired assertion, that $\psi: M^{(r)} \longrightarrow J(M)$ is a locally trivial analytic fibration over $J(M)$ with fibre $\mathbb{P}^{r-g}$.

It should be noted that the natural complement to Theorem 14(a) also follows readily from these observations.

Theorem 14(b). If V is an irreducible component of the analytic subvariety $W_r^\nu \subseteq J(M)$ and if V is not entirely contained within $W_r^{\nu+1}$ then

$$\dim V \geqq r\nu - (\nu-1)(g+\nu) .$$

Proof. If $V_o \subseteq V$ is a sufficiently small open subset of V, and $V_o \cap W_r^{\nu+1} = \emptyset$, then it follows from Theorem 15 that there is a one-one analytic mapping $\lambda_o: V_o \times \mathbb{P}^{\nu-1} \longrightarrow \psi^{-1}(V_o) \subseteq M^{(r)}$; and consequently $\dim \psi^{-1}(V_o) = \dim V + \nu-1$. On the other hand, $\psi^{-1}(V_o)$ is an open subset of G_r^ν not contained in $G_r^{\nu+1}$, hence it follows from Theorem 14(a) that $\dim \psi^{-1}(V_o) \geqq r\nu - (\nu-1)(g+\nu-1)$. The desired result follows immediately from these two formulas, and the proof is thereby concluded.

As a brief digression, it is of some interest to examine more closely the Jacobi mapping $\psi: M^{(g)} \longrightarrow J(M)$, to secure a more detailed analysis of the solution of the Jacobi inversion problem; this can be done rather completely for surfaces of sufficiently small genus. Note first that as a consequence of Corollary 3 to Theorem 10' the Jacobi mapping induces a complex analytic homeomorphism

$$\psi: M^{(g)} \backslash G_g^2 \longrightarrow J(M) \backslash W_g^2 .$$

It follows from the Riemann-Roch theorem as rewritten in the form of equation (13) of §2 that $W_g^2 = k - W_{g-2}$, where $k \in J(M)$ is the canonical point; thus W_g^2 is an irreducible complex analytic subvariety of $J(M)$ of dimension $g-2$. It then further follows from Theorem 11(a) that the singular locus of the subvariety W_{g-2} is $\mathcal{S}(W_{g-2}) = W^2_{g-2}$; hence applying the same form of the Riemann-Roch theorem as above, $\mathcal{S}(W^2_g) = k - W^2_{g-2} = W^3_g$. Over the regular locus $\mathcal{R}(W^2_g) = W^2_g \backslash W^3_g$ the Jacobi mapping

$$\psi\colon G^2_g \backslash G^3_g \longrightarrow W^2_g \backslash W^3_g$$

is a locally trivial analytic fibration with fibre $\mathbb{P}^1$, as a consequence of Corollary 1 to Theorem 15. Recall from Theorem 7 that $\dim W^3_g \leqq g-4$, hence that $W^3_g = \emptyset$ when $g = 2,3$; and recall from Theorem 8 that $W^3_g = \emptyset$ when $g = 4$ and the surface M is not hyperelliptic, while $W^3_g = -2e$ when $g = 4$ and the surface M is hyperelliptic, where $e \in J(M)$ is the hyperelliptic point.

Thus if M is a Riemann surface of genus $g = 2$, so that M is hyperelliptic and $W^2_g = e$ is the hyperelliptic point of $J(M)$, the Jacobi mapping $\psi\colon M^{(2)} \longrightarrow J(M)$ has the properties that $\psi^{-1}(e)$ is an analytic submanifold of $M^{(2)}$ analytically homeomorphic to $\mathbb{P}^1$, and that $\psi\colon M^{(2)} \backslash \psi^{-1}(e) \longrightarrow J(M) \backslash e$ is a complex analytic homeomorphism; the symmetric product $M^{(2)}$ is obtained from the Jacobi variety merely by blowing up the hyperelliptic point $e \in J(M)$ to a projective line $\mathbb{P}^1$, using the picturesque terminology that has been introduced in complex analysis and algebraic geometry to describe such mappings. If M is a Riemann surface of genus $g = 3$ then $W^2_g = k - W_1$ is an analytic submanifold of $J(M)$ analytically homeomorphic to the surface M

itself; the Jacobi mapping $\psi: M^{(3)} \longrightarrow J(M)$ has the properties that $\psi: G_g^2 \longrightarrow W_g^2$ is a locally trivial analytic fibration over the manifold W_g^2 with fibre $\mathbb{P}^1$, and that $\psi: M^{(3)} \setminus G_g^2 \longrightarrow J(M) \setminus W_g^2$ is a complex analytic homeomorphism. The symmetric product $M^{(3)}$ is obtained from the Jacobi variety by blowing up each point of the analytic submanifold $W_g^2 \subset J(M)$ to a projective line. If M is a Riemann surface of genus $g = 4$ then $W_g^2 = k - W_2$ is an analytic subvariety of $J(M)$ which is the image of an analytic mapping $\psi^* = k - \psi: M^{(2)} \longrightarrow W_g^2 \subset J(M)$; if M is not hyperelliptic this mapping is a complex analytic homeomorphism and its image is a regularly imbedded analytic submanifold of $J(M)$, while if M is hyperelliptic the image W_g^2 has an isolated singularity at the point $-2e \in W_g^2 \subset J(M)$ and the analytic mapping ψ^* exhibits the manifold $M^{(2)}$ as being obtained from the analytic variety W_g^2 by blowing this singular point up into a projective line. The Jacobi mapping $\psi: M^{(4)} \longrightarrow J(M)$ has the properties that $\psi: G_g^2 \setminus G_g^3 \longrightarrow \mathcal{R}(W_g^2)$ is a locally trivial analytic fibration over the regular locus of the subvariety $W_g^2 \subset J(M)$ with fibre $\mathbb{P}^1$, that the isolated singular point of W_g^2 (if it exists) has as inverse image $\psi^{-1}(-2e) = G_g^3$ an analytic submanifold of $M^{(4)}$ analytically homeomorphic to $\mathbb{P}^2$, and that $\psi: M^{(4)} \setminus G_g^2 \longrightarrow J(M) \setminus W_g^2$ is a complex analytic homeomorphism. The symmetric product $M^{(4)}$ is obtained from the Jacobi variety by blowing up each point of the analytic submanifold $\mathcal{R}(W_g^2) \subset J(M)$ to a projective line, and blowing up the isolated singular point of W_g^2 (if it exists) to a projective plane; the singularity occurs only when M is hyperelliptic.

(g) The proof of Theorem 15 was really accomplished by demonstrating the existence of an analytic mapping $\mu_V\colon V\times \mathbf{C}^\nu \longrightarrow \psi^{-1}(V)$, over any sufficiently small relatively open subset $V \subseteq W_r^\nu \backslash W_r^{\nu+1}$, that factors through the natural mapping $\mathbf{C}^\nu \longrightarrow \mathbb{P}^{\nu-1}$ to yield the desired analytic mapping $\lambda_V\colon V \times \mathbb{P}^{\nu-1} \longrightarrow \psi^{-1}(V)$; indeed the mapping μ_V constructed in that proof can evidently be viewed as the composite $\mu_V = \mu_V''\mu_V'$ of a one-one mapping

$$\mu_V'\colon V \times \mathbf{C}^\nu \longrightarrow \bigcup_{\xi\in V} \Gamma(M,\mathcal{O}(\xi\zeta_{p_o}^r))$$

which for each $\xi \in V$ is a linear isomorphism

$$\mu_V'(\xi)\colon \mathbf{C}^\nu \longrightarrow \Gamma(M,\mathcal{O}(\xi\zeta_{p_o}^r)) ,$$

and the divisor mapping

$$\mu_V''\colon \bigcup_{\xi\in V} \Gamma(M,\mathcal{O}(\xi\zeta_{p_o}^r)) \longrightarrow \psi^{-1}(V)$$

which for any $\xi \in V$ and $f \in \Gamma(M,\mathcal{O}(\xi\zeta_{p_o}^r))$ has the value

$$\mu_V''(\xi,f) = \mathscr{D}(f) \in M^{(r)} .$$

The mapping μ_V' can be thought of as determining on the set $\bigcup_{\xi\in V} \Gamma(M,\mathcal{O}(\xi\zeta_{p_o}^r))$ the structure of a complex analytic variety $V \times \mathbf{C}^\nu$, or equivalently, as determining on the set $\bigcup_{\xi\in V} \Gamma(M,\mathcal{O}(\xi\zeta_{p_o}^r))$

the structure of a complex analytic vector bundle over the analytic variety V ; and the mapping μ''_V is then a one-one analytic mapping from the complex analytic projective bundle naturally associated to this vector bundle onto the analytic variety $\psi^{-1}(V)$. It is quite natural to expect that this local construction can be extended to a global construction, that is, that the set $\bigcup_{\xi \in W_r^\nu \setminus W_r^{\nu+1}} \Gamma(M, \mathcal{O}(\xi\zeta_{p_0}^r))$ can be given the natural structure of a complex analytic vector bundle over the variety $W_r^\nu \setminus W_r^{\nu+1}$ such that the divisor mapping is a one-one analytic mapping from the complex analytic projective bundle naturally associated to this vector bundle onto the analytic variety $G_r^\nu \setminus G_r^{\nu+1}$; this is of course equivalent to asserting merely that the local mappings μ'_V for overlapping open sets V are compatible, that is, that whenever V_1, V_2 are intersecting open subsets of $W_r^\nu \setminus W_r^{\nu+1}$ for which this construction has been carried out there exists a complex analytic homeomorphism

$$\mu_{V_1V_2}: (V_1 \cap V_2) \times \mathbf{C}^\nu \longrightarrow (V_1 \cap V_2) \times \mathbf{C}^\nu$$

which for each fixed point $\xi \in V_1 \cap V_2$ is linear on the space $\mathbf{C}^\nu$ and which is such that $\mu'_{V_1}\mu_{V_1V_2} = \mu'_{V_2}$ on $(V_1 \cap V_2) \times \mathbf{C}^\nu$. The mappings $\mu_{V_1V_2}$ are then the coordinate transition functions for the vector bundle so defined on $W_r^\nu \setminus W_r^{\nu+1}$. In demonstrating that this global assertion is indeed true it is convenient, and for some other applications quite useful, to show something rather more precise.

Any analytic line bundle $\xi \in P(M) \subset H^1(M, \mathcal{O}^*)$ can be represented by a flat line bundle $X \in H^1(M,\mathbf{C}^*)$, as observed in §8(a) of the earlier lecture notes. Indeed the set of flat line bundles over a Riemann surface M of genus g form a complex Lie group isomorphic to $(\mathbf{C}^*)^{2g}$, the subset of analytically trivial flat line bundles form a Lie subgroup isomorphic to $\mathbf{C}^g$, and the quotient group is isomorphic to $P(M)$, so that the manifold $H^1(M,\mathbf{C}^*)$ is a complex analytic principal bundle over the manifold $P(M)$ with group and fibre equal to $\mathbf{C}^g$; this is of course merely an interpretation of the exact cohomology sequence

$$(8) \qquad 0 \longrightarrow \Gamma(M,\mathcal{O}^{1,0}) \xrightarrow{\delta^*} H^1(M,\mathbf{C}^*) \xrightarrow{i^*} P(M) \longrightarrow 0$$

derived on page 132 of the earlier lecture notes. It is further evident from the proof of Lemma 27 of the earlier lecture notes that the flat line bundles over M can all be represented by cocycles $\{\chi_{\alpha\beta}\} \in Z^1(\mathcal{U},\mathbf{C}^*)$ for a fixed open covering $\mathcal{U} = \{U_\alpha\}$ of the Riemann surface, in such a manner that these cocycles are complex analytic functions $\chi_{\alpha\beta}(X)$ of the points $X \in H^1(M,\mathbf{C}^*)$. It is then an immediate consequence of these observations that for any sufficiently small open subset $V \subseteq P(M)$ and any index r the line bundles $\xi\zeta_{p_o}^r$ for all $\xi \in V$ can be represented by cocycles $\{\xi_{\alpha\beta}\} \in Z^1(\mathcal{U}, \mathcal{O}^*)$ for a fixed open covering $\mathcal{U} = \{U_\alpha\}$ of the Riemann surface, and in such a manner that the cocycles $\xi_{\alpha\beta} = \xi_{\alpha\beta}(\xi,z)$ are complex analytic functions of $\xi \in V \subseteq P(M)$ and of $z \in U_\alpha \cap U_\beta \subseteq M$; for the analytic fibration

$H^1(M,\mathbf{C}^*) \longrightarrow P(M)$ admits local sections. This provides a very convenient explicit representation of the local bundle structure on the sets $\bigcup_{\xi \in V} \Gamma(M, \mathcal{O}(\xi\zeta^r_{p_0}))$.

Theorem 16(a). For any sufficiently small relatively open subset V of the analytic subvariety $W^\nu_r \setminus W^{\nu+1}_r \subseteq P(M)$ and any index r, there exist an open covering $\mathcal{U} = \{U_\alpha\}$ of the Riemann surface M and complex analytic mappings $\xi_{\alpha\beta}\colon V \times (U_\alpha \cap U_\beta) \longrightarrow \mathbf{C}^*$ so that for each fixed $\xi \in V$ the functions $\{\xi_{\alpha\beta}(\xi)\} \in Z^1(\mathcal{U}, \mathcal{O}^*)$ form a cocycle representing the line bundle $\xi\zeta^r_{p_0}$; and furthermore, there exist complex analytic mappings $f^i_\alpha\colon V \times U_\alpha \longrightarrow \mathbf{C}$, $1 \le i \le \nu$, so that for each fixed $\xi \in V$ the functions $f^i = \{f^i_\alpha(\xi)\} \in \Gamma(M, \mathcal{O}(\xi\zeta^r_{p_0}))$ form a basis for the space of holomorphic sections of the line bundle $\xi\zeta^r_{p_0}$, $1 \le i \le \nu$, when that bundle is represented by the cocycle $\{\xi_{\alpha\beta}(\xi)\} \in Z^1(\mathcal{U}, \mathcal{O}^*)$.

Proof. Although the first assertion of the theorem was proved just above, it is convenient to prove both assertions simultaneously; but the preceding proof may well serve as an enlightening motivation for the constructions in this proof. The proof is really a natural continuation of the proof of Theorem 15, so the notation and terminology introduced in the course of that proof will be presupposed here. To continue, then, introduce the universal covering space $\tilde{M}$ of the Riemann surface M; and assuming that the open sets $U_j \subseteq M$, $1 \le j \le g$, are simply connected, select open sets $\tilde{U}_j \subseteq \tilde{M}$ covering each of the sets U_j simply, and let $z_j \in \tilde{U}_j$ denote the point lying over $p_j \in U_j$ and $z^0_j \in \tilde{U}_j$

denote the point lying over p_j^o. It can be assumed that the base point $p_o \in M$ is not contained in any of the sets U_j. In terms of the prime function of the marked Riemann surface M, define

$$g(z_1,\ldots,z_g;z) = \prod_{j=1}^{g} p(z,z_o;z_j^o,z_j) .$$

This is then a meromorphic function on $\tilde{U}_1 \times \ldots \times \tilde{U}_g \times \tilde{M}$, which is identically 1 when $z = z_o$, and which as a function of $z \in \tilde{M}$ has simple zeros at the points Γz_j^o and simple poles at the points Γz_j, and is otherwise holomorphic and nonvanishing on $\tilde{M}$; of course when $z_j = z_j^o$ this function remains holomorphic and nonvanishing at that point also. Furthermore, for any transformation T belonging to the covering translation group Γ, this function has the property that

$$g(z_1,\ldots,z_g;Tz) = \chi(T;z_1,\ldots,z_g)g(z_1,\ldots,z_g;z) ,$$

where $\chi(T;z_1,\ldots,z_g)$ is a holomorphic nonvanishing function on $\tilde{U}_1 \times \ldots \times \tilde{U}_g$ for each $T \in \Gamma$; the explicit form of this function can readily be determined also, referring back to Theorem 4, but will not be needed here. The functions $\chi(T;z_1,\ldots,z_g)$ really describe flat line bundles over M depending analytically on the parameters $(z_1,\ldots,z_g) \in \tilde{U}_1 \times \ldots \times \tilde{U}_g$, or equivalently of course, depending analytically on the parameters $(p_1,\ldots,p_g) \in U_1 \times \ldots \times U_g$; and the functions $g(z_1,\ldots,z_g;z)$ correspondingly describe a meromorphic family of meromorphic sections of these bundles. In more detail, for any covering $\mathcal{U} = \{U_\alpha\}$ of the Riemann surface M by simply connected open subsets $U_\alpha \subset M$ with connected pairwise intersections, select for each set $U_\alpha \in \mathcal{U}$ an open subset $\tilde{U}_\alpha \subset \tilde{M}$ simply covering U_α ; and note that for each nonempty intersection $U_\alpha \cap U_\beta$ there is a unique covering translation $T_{\alpha\beta} \in \Gamma$ such that $\tilde{U}_\alpha \cap T_{\alpha\beta}\tilde{U}_\beta \neq \emptyset$, namely, that transformation $T_{\alpha\beta}$ taking the portion of $\tilde{U}_\beta$ lying over $U_\alpha \cap U_\beta$ to the portion of $\tilde{U}_\alpha$ lying over $U_\alpha \cap U_\beta$. Then to each nonempty intersection $U_\alpha \cap U_\beta$ associate the function $\chi_{\alpha\beta}(p_1,\ldots,p_g) = \chi(T_{\alpha\beta};z_1,\ldots,z_g)$; these are complex analytic mappings $\chi_{\alpha\beta}\colon U_1 \times\ldots\times U_g \longrightarrow \mathbf{C}^*$, which for each fixed point $(p_1,\ldots,p_g) \in U_1 \times\ldots\times U_g$ are easily seen to form a cocycle

$$\chi(p_1,\ldots,p_g) = \{\chi_{\alpha\beta}(p_1,\ldots,p_g)\} \in Z^1(\mathcal{U},\mathbf{C}^*)$$

and hence to define a flat line bundle over M . Also to each set $U_\alpha \in \mathcal{U}$ associate the function $g_\alpha(p_1,\ldots,p_g;p) = g(z_1,\ldots,z_g;z_\alpha)$ where $z_\alpha \in \tilde{U}_\alpha$ lies over $p \in U_\alpha$; these are meromorphic functions $g_\alpha\colon U_1 \times\ldots\times U_g \times U_\alpha \longrightarrow \mathbf{C}$ which for each fixed point $(p_1,\ldots,p_g) \in U_1 \times\ldots\times U_g$ are easily seen to form a meromorphic section

$$g(p_1,\ldots,p_g) = \{g_\alpha(p_1,\ldots,p_g;p)\} \in \Gamma(M,\mathcal{M}(\chi(p_1,\ldots,p_g)))$$

of the flat line bundle $\chi(p_1,\ldots,p_g)$. Note that by construction $\mathcal{Y}(g(p_1,\ldots,p_g)) = p_1^o +\ldots+ p_g^o - p_1 - \ldots - p_g$; and consequently as complex analytic line bundles, $\chi(p_1,\ldots,p_g) = \zeta_{\mathcal{Y}}\zeta_{p_1}^{-1}\cdots\zeta_{p_g}^{-1} \in H^1(M,\mathcal{O}$

With this auxiliary construction out of the way, the remainder of the proof then follows readily from the proof of Theorem 15.

Suppose that the line bundles η, ζ_{ϑ}, and $\zeta_{p_0}^{r}$ are also defined by analytic cocycles in $Z^1(\mathfrak{U}, \mathcal{O}^*)$, for the same covering $\mathfrak{U}$ of the Riemann surface M; the products of the appropriate cocycles then clearly yield complex analytic mappings

$$\xi_{\alpha\beta}: U_1 \times \ldots \times U_g \times (U_\alpha \cap U_\beta) \longrightarrow \mathbf{C}^*$$

which for each fixed point $(p_1, \ldots, p_g) \in U_1 \times \ldots \times U_g$ form a cocycle $\xi = \{\xi_{\alpha\beta}(p_1, \ldots, p_g; p)\} \in Z^1(\mathfrak{U}, \mathcal{O}^*)$ representing the complex analytic line bundle $\xi\zeta_{p_0}^{r} = \eta\chi\zeta_{p_0}^{r} = \eta\zeta_{\vartheta}\zeta_{p_1}^{-1} \cdots \zeta_{p_g}^{-1}\zeta_{p_0}^{r} =$ $= \eta\zeta_{\vartheta}\zeta_{p_0}^{r} \cdot \zeta_{\vartheta}\zeta_{p_1}^{-1} \cdots \zeta_{p_g}^{-1} \cdot \zeta_{\vartheta}^{-1}$, where $\xi = \eta\chi \in P(M)$. The restrictions of these functions to the subvariety $V \subseteq U_1 \times \ldots \times U_g$ representing W_r^{ν} in a neighborhood of the point $\eta \in W_r^{\nu} \backslash W_r^{\nu+1}$ can be viwed as analytic mappings $\xi_{\alpha\beta}: V \times (U_\alpha \cap U_\beta) \longrightarrow \mathbf{C}^*$ having the properties required by the first assertion of the theorem. The sections $f_j \in \Gamma(M, \mathcal{O}(\eta\zeta_{\vartheta}\zeta_{p_0}^{r}))$, $1 \leq j \leq n$, forming a basis for the space of sections of the line bundle $\eta\zeta_{\vartheta}\zeta_{p_0}^{r}$ can be represented by local complex analytic maps $f_{j\alpha}: U_\alpha \longrightarrow \mathbf{C}$, when the bundle $\eta\zeta_{\vartheta}\zeta_{p_0}^{r}$ is represented by a cocycle in $Z^1(\mathfrak{U}, \mathcal{O}^*)$; and there are complex analytic mappings $h_\alpha: U_\alpha \longrightarrow \mathbf{C}$ representing a section $h = \{h_\alpha\} \in \Gamma(M, \mathcal{O}(\zeta_{\vartheta}))$ of the line bundle ζ_{ϑ} when that bundle is represented by a cocycle in $Z^1(\mathfrak{U}, \mathcal{O}^*)$, such that $\vartheta(h) = \vartheta$. Then using the analytic mappings $c_j^i: V \longrightarrow \mathbf{C}$ introduced in the proof of Theorem 15, define the meromorphic functions f_α^i on $V \times U_\alpha$ by

$$f^i_\alpha(p_1,\ldots,p_g;p) = \sum_{j=1}^{n} c^i_j(p_1,\ldots,p_g)\, f_{j\alpha}(p)\, g_\alpha(p_1,\ldots,p_g;p)\, h_\alpha(p)^{-1} .$$

Note that for each point $(p_1,\ldots,p_g) \in V$ it follows from the constructions that the functions $f^i = \{f^i_\alpha(p_1,\ldots,p_g;p)\}$ are linearly independent meromorphic sections of the line bundle $\xi\zeta^r_{p_o} = \eta\zeta_{\mathscr{J}}\,\zeta^r_{p_o}\cdot\zeta_{\mathscr{J}}\,\zeta^{-1}_{p_1}\cdots\zeta^{-1}_{p_g}\cdot\zeta^{-1}_{\mathscr{J}}$; and since in addition $\mathscr{J}(\Sigma_j\, c^i_j(p_1,\ldots,p_g)f_j) \geqq p_1 +\ldots+ p_g$, $\mathscr{J}(g(p_1,\ldots,p_g)) = \mathscr{J} - p_1 - \ldots -$ and $\mathscr{J}(h^{-1}) = -\mathscr{J}$, these sections are really holomorphic, hence form a basis for the space of holomorphic sections of the line bundle $\xi\zeta^r_{p_o}$. That serves to conclude the proof of the theorem.

Corollary 1 to Theorem 16. In the conclusions of Theorem 16(a) the mappings $\xi_{\alpha\beta}\colon V \times (U_\alpha \cap U_\beta) \longrightarrow \mathbf{C}^*$ can be taken to be of the form $\xi_{\alpha\beta}(\xi,p) = \chi_{\alpha\beta}(\xi)\zeta_{\alpha\beta}(p)$, where $\chi_{\alpha\beta}\colon V \longrightarrow \mathbf{C}^*$ are analytic mappings which for each fixed point $\xi \in V$ form a cocycle representing a flat line bundle $\chi(\xi) \in H^1(M,\mathbf{C}^*)$ which in turn represents the complex analytic line bundle ξ , and $\zeta_{\alpha\beta}\colon U_\alpha \cap U_\beta \longrightarrow \mathbf{C}^*$ are analytic mappings forming a cocycle representing the complex analytic line bundle $\zeta^r_{p_o}$.

Proof. This is merely an observation of what was actually proved in the course of the proof of Theorem 16(a).

Applying this theorem, for any sufficiently small relatively open subset V of the analytic subvariety $W^\nu_r \backslash W^{\nu+1}_r \subseteq P(M)$ and any index r , introduce the mapping

$$\mu_V\colon V \times \mathbf{C}^\nu \longrightarrow \bigcup_{\xi \in V} \Gamma(M, \mathcal{O}(\xi\zeta^r_{p_o}))$$

defined by

$$\mu_V(\xi;t_1,\dots,t_\nu) = \sum_{i=1}^{\nu} t_i f^i(\xi)$$

where the functions $f^i(\xi) = \{f^i_\alpha(\xi)\} \in \Gamma(M, \mathcal{O}(\xi\zeta^r_{p_o}))$ are as in the statement of Theorem 16(a). This mapping is clearly one-to-one and linear for each fixed $\xi \in V$, hence establishes the local product structure or equivalently the vector bundle structure desired on the set $\bigcup_{\xi \in V} \Gamma(M, \mathcal{O}(\xi\zeta^r_{p_o}))$, in a quite explicit form; indeed the form is sufficiently explicit that this bundle structure is easily extended to the entire set $\bigcup_{\xi \in W^\nu_r \setminus W^{\nu+1}_r} \Gamma(M, \mathcal{O}(\xi\zeta^r_{p_o}))$.

Let $\{V_A\}$ be a covering of the analytic subvariety $W^\nu_r \setminus W^{\nu+1}_r \subseteq P(M)$ by sufficiently small relatively open subsets such that the constructions of Theorem 16(a) and of its Corollary 1 can be carried out on each of the sets V_A. Note that these constructions can be carried out in terms of the same open covering $\mathfrak{U} = \{U_\alpha\}$ of the Riemann surface M for all the sets V_A simultaneously, even if there are not necessarily only finitely many of these sets V_A; for all the flat line bundles over M as well as the bundles ζ_ϑ and $\zeta^r_{p_o}$ can be described in terms of one covering $\mathfrak{U}$, which can be assumed to consist of simply connected open subsets $U_\alpha \subset M$ with connected pairwise intersections. Thus there are complex analytic mappings $\chi_{A\alpha\beta}: V_A \longrightarrow \mathbf{C}^*$ and $\zeta_{\alpha\beta}: U_\alpha \cap U_\beta \longrightarrow \mathbf{C}^*$ such that for any fixed point $\xi \in V_A$ the functions $\{\chi_{A\alpha\beta}(\xi)\zeta_{\alpha\beta}\} \in Z^1(\mathfrak{U}, \mathcal{O}^*)$ form a cocycle representing the complex analytic line bundle $\xi\zeta^r_{p_o}$; and there are complex analytic mappings $f^i_{A\alpha}: V_A \times U_\alpha \longrightarrow \mathbf{C}$, $1 \leq i \leq \nu$, so that for any fixed point $\xi \in V_A$ the functions $\{f^i_{A\alpha}(\xi)\} \in \Gamma(M, \mathcal{O}(\xi\zeta^r_{p_o}))$ are a basis for the space of holomorphic sections of the line bundle $\xi\zeta^r_{p_o}$. The condition that these functions are sections of the line bundle means explicitly that

$$f^i_{A\alpha}(\xi,p) = \chi_{A\alpha\beta}(\xi)\zeta_{\alpha\beta}(p)f^i_{A\beta}(\xi,p) \tag{9}$$

whenever $\xi \in V_A$ and $p \in U_\alpha \cap U_\beta$. Then over each of these sets V_A the family $\bigcup_{\xi \in V_A} \Gamma(M, \mathcal{O}(\xi\zeta^r_{p_o}))$ can be given a local product structure by the mapping

$$\mu_A: V_A \times \mathbf{C}^\nu \longrightarrow \bigcup_{\xi \in V_A} \Gamma(M, \mathcal{O}(\xi\zeta^r_{p_o}))$$

defined by

$$\mu_A(\xi;t^A_1,\ldots,t^A_\nu) = \{\sum_{i=1}^{\nu} t^A_i f^i_A(\xi)\} \ ; \tag{10}$$

the constants $(t^A_1,\ldots,t^A_\nu)$ are thus the fibre coordinates of the resulting vector bundle over the coordinate neighborhood $V_A \subseteq W^\nu_r \backslash W^{\nu+1}_r$, and it merely remains to compare these coordinatizations over intersections $V_A \cap V_B$ of coordinate neighborhoods.

For any point $\xi \in V_A \cap V_B$, the cocycles $\chi_A(\xi) = \{\chi_{A\alpha\beta}(\xi)\}$ and $\chi_B(\xi) = \{\chi_{B\alpha\beta}(\xi)\}$ in $Z^1(\mathfrak{U},\mathbf{C}^*)$ determine flat line bundles over M which represent the same complex analytic line bundle $\xi\zeta^r_{p_o}$; the functions χ_A and χ_B can therefore be viewed as two sections over $V_A \cap V_B$ of the complex analytic bundle

$H^1(M,\mathbf{C}^*) \longrightarrow P(M)$ described by the exact sequence (8). Thus there exists a complex analytic mapping $\sigma_{AB}: V_A \cap V_B \longrightarrow \Gamma(M, \mathcal{O}^{1,0})$ so that

(11) $$\chi_A(\xi) = \delta^* \sigma_{AB}(\xi) \cdot \chi_B(\xi) \quad \text{for all} \quad \xi \in V_A \cap V_B .$$

To be more explicit, in terms of the basis $\{\omega_i\}$ for the space of Abelian differentials over M, there are complex analytic mappings $s_{ABi}: V_A \cap V_B \longrightarrow \mathbf{C}$ such that $\sigma_{AB}(\xi) = \sum_{i=1}^{g} s_{ABi}(\xi)\omega_i$. For each open set U_α select analytic functions $w_{i\alpha}$ in U_α such that $\omega_i = dw_{i\alpha}$ in U_α; and recalling the explicit form for the coboundary mapping δ^* in the exact sequence (8), it follows that (11) can be rewritten

(12) $$\chi_{A\alpha\beta}(\xi) = \exp\{2\pi i \sum_{i=1}^{g} s_{ABi}(\xi)[w_{i\beta}(p) - w_{i\alpha}(p)]\} \cdot \chi_{B\alpha\beta}(\xi)$$

for all $\xi \in V_A \cap V_B$ and $p \in U_\alpha \cap U_\beta$. Then introducing the complex analytic mappings $s_{AB\alpha}: (V_A \cap V_B) \times U_\alpha \longrightarrow \mathbf{C}^*$ defined by $s_{AB\alpha}(\xi,p) = \exp\{-2\pi i \sum_{i=1}^{g} s_{ABi}(\xi)w_{i\alpha}(p)\}$, condition (12) can finally be rewritten in the form

(13) $$\chi_{A\alpha\beta}(\xi) = \frac{s_{AB\alpha}(\xi,p)}{s_{AB\beta}(\xi,p)} \chi_{B\alpha\beta}(\xi)$$

for all $\xi \in V_A \cap V_B$ and $p \in U_\alpha \cap U_\beta$. Now for any $\xi \in V_A \cap V_B$ the functions $\{f^i_{A\alpha}(\xi)\}$, $1 \leq i \leq \nu$, represent a basis for the space of holomorphic sections of the line bundle $\xi\zeta^r_{p_0}$, as do the functions $\{f^i_{B\alpha}(\xi)\}$, $1 \leq i \leq \nu$; but these represent holomorphic sections in terms of two different cocycles describing the line bundle $\xi\zeta^r_{p_0}$, in the sense that the functions $\{f^i_{A\alpha}\}$ satisfy equation (9) while the functions $\{f^i_{B\alpha}\}$ satisfy the corresponding equation over V_B. However in view of (13) the functions $\{s_{AB\alpha}(\xi,p)f^i_{B\alpha}(\xi,p)\}$ also satisfy (9), hence represent with the functions $\{f^i_{A\alpha}(\xi,p)\}$ two bases for the space of sections of the line bundle $\xi\zeta^r_{p_0}$, expressed in terms of the same cocycle representation for that line bundle; thus for any $\xi \in V_A \cap V_B$ there are constants $\{\chi^{ij}_{AB}(\xi)\}$ determining a nonsingular matrix $\chi_{AB}(\xi) \in GL(\nu,\mathbf{C})$ such that

$$(14) \qquad f^i_{A\alpha}(\xi,p) = \sum_{j=1}^{\nu} \chi^{ij}_{AB}(\xi)s_{AB\alpha}(\xi,p)f^j_{B\alpha}(\xi,p)$$

whenever $\xi \in V_A \cap V_B$ and $p \in U_\alpha$. Since the sections $\{f^i_{A\alpha}(\xi)\}$ are linearly independent, then for any point $\xi \in V_A \cap V_B$ there are points $p_k \in U_\alpha$ such that the matrix $\{f^i_{B\alpha}(\xi,p_k)\}$, $i,k = 1,\ldots,\nu$, is nonsingular at ξ and hence in an open neighborhood of ξ; and it follows immediately from (14) that the functions $\chi^{ij}_{AB}(\xi)$ are holomorphic in ξ, since they are uniquely determined by the equations (14) at the ν points p_k. Thus considering merely the sections $f^i_A(\xi)$, $f^i_B(\xi)$ in $\Gamma(M,\mathcal{O}(\xi\zeta^r_{p_0}))$ represented by the functions $\{f^i_{A\alpha}(\xi,p)\}$, $\{f^i_{B\alpha}(\xi,p)\}$, it follows from (14) that

$$(15) \qquad f^i_A(\xi) = \sum_{j=1}^{\nu} \chi^{ij}_{AB}(\xi)f^i_B(\xi) \quad \text{whenever} \quad \xi \in V_A \cap V_B ,$$

for the complex analytic mappings $\chi_{AB}\colon V_A \cap V_B \longrightarrow GL(\nu,\mathbf{C})$;

the cocycle $\{\chi_{AB}\} \in Z^1(\{V_A\},\mathcal{O}(GL(\nu,\mathbf{C})))$ then describes the coordinate transformations relating the coordinatizations (10) over intersecting neighborhoods V_A and V_B, in the sense that $\mu_A(\xi;t_1^A,\ldots,t_\nu^A) = \mu_B(\xi;t_1^B,\ldots,t_\nu^B)$ for a point $\xi \in V_A \cap V_B$ if and only if

$$t_i^B = \sum_{j=1}^{\nu} t_j^A \chi_{AB}^{ji}(\xi) \; . \tag{16}$$

These observations taken together with the preceding results then yield immediately the proof of the following assertion.

Theorem 16(b). The set $\bigcup_{\xi \in W_r^\nu \setminus W_r^{\nu+1}} \Gamma(M,\mathcal{O}(\xi\zeta_{p_o}^r))$ can be given the structure of a complex analytic vector bundle over the analytic variety $W_r^\nu \setminus W_r^{\nu+1}$, in such a manner that the fibre over $\xi \in W_r^\nu \setminus W_r^{\nu+1}$ is the vector space $\Gamma(M,\mathcal{O}(\xi\zeta_{p_o}^r))$; and the divisor mapping

$$\vartheta : \Gamma(M,\mathcal{O}(\xi\zeta_{p_o}^r)) \longrightarrow M^{(r)}$$

induces a one-one complex analytic mapping from the associated complex analytic projective bundle over $W_r^\nu \setminus W_r^{\nu+1}$ with fibre $\mathbb{P}^{\nu-1}$ onto the subvariety $G_r^\nu \setminus G_r^{\nu+1} \subseteq M^{(r)}$, so that $\psi\vartheta$ is the bundle projection.

In the special case $r \geqq 2g-1$ there follows immediately the following natural complement of Corollary 2 to Theorem 10 and Corollary 2 to Theorem 15.

<u>Corollary 2 to Theorem 16.</u> For any index $r \geqq 2g-1$ the set $\bigcup_{\xi \in J(M)} \Gamma(M, \mathcal{O}(\xi\zeta_{p_o}^r))$ can be given the structure of a complex analytic vector bundle over the manifold $J(M)$, in such a manner that the fibre over $\xi \in J(M)$ is the vector space $\Gamma(M, \mathcal{O}(\xi\zeta_{p_o}^r))$; and the divisor mapping

$$\mathcal{J} : \Gamma(M, \mathcal{O}(\xi\zeta_{p_o}^r)) \longrightarrow M^{(r)}$$

induces an analytic fibre bundle homeomorphism between the associated complex analytic projective bundle and the bundle $\psi: M^{(r)} \longrightarrow J(M)$.

Notes for §3.

(a) An analytic mapping $\tau: M \longrightarrow N$ between two complex analytic manifolds, with the property that the inverse image of any point of N is a nonempty finite set of points of M, is locally a branched analytic covering, in the technical sense customary in the study of complex analytic varieties; that is to say, the mapping τ is proper and light, and there exists a proper complex analytic subvariety $D \subset N$ for which the restriction $\tau: M - \tau^{-1}(D) \longrightarrow N - D$ is a complex analytic covering projection. A discussion of the properties of such mappings can be found in most texts on functions of several complex variables; the terminology used here is that of R. C. Gunning, Lectures on Complex Analytic Varieties (Princeton University Press, 1970). In particular, whenever τ is a one-one mapping it is an analytic homeomorphism. Similar mappings can also be considered where both domain and range are complex analytic varieties with possible singularities, although then they need not be analytic homeomorphisms even when one-one; but such mappings do always preserve dimension.

(b) The subvarieties $W_r - W_r \subseteq J(M)$ have a definite interest, as evidenced for instance in the course of the proof of Theorem 11. They are irreducible analytic subvarieties of $J(M)$, and $\dim(W_r - W_r) = 2r$ whenever $2r \leqq g$; and the complex analytic mapping $\Psi: M^{(r)} \times M^{(r)} \longrightarrow J(M)$ defined by $\Psi(\mathfrak{Y}_1, \mathfrak{Y}_2) = \psi(\mathfrak{Y}_1) - \psi(\mathfrak{Y}_2)$ has as image precisely the subvariety

$W_r - W_r$.

It is quite easy to derive the analogue of Theorem 10 for this mapping Ψ , particularly for the case $r = 1$. Considering the mapping $\Psi: M \times M \longrightarrow J(M)$, note that $\Psi(p_1,q_1) = \Psi(p_2,q_2)$ precisely when $\varphi(p_1) + \varphi(q_2) = \varphi(p_2) + \varphi(q_1)$, hence precisely when $p_1 + q_2 \approx p_2 + q_1$. If the latter two divisors coincide and the points (p_1,q_1) and (p_2,q_2) are distinct then necessarily $p_1 = q_1$ and $p_2 = q_2$, while if the two divisors are distinct but linearly equivalent they must be the divisors of two separate sections of a line bundle in W_2^2 and the Riemann surface is necessarily hyperelliptic; indeed in this second case necessarily $p_2 = \theta q_1$ and $q_2 = \theta p_2$, where $\theta: M \longrightarrow M$ is the hyperelliptic involution. Thus if M is not hyperelliptic the mapping $\Psi: M \times M \longrightarrow W_1 - W_1$ maps the diagonal $D = \{(p,p) \in M \times M\}$ to zero and is otherwise a one-one complex analytic mapping $\Psi: (M \times M)\setminus D \longrightarrow (W_1 - W_1)\setminus 0$; while if M is hyperelliptic with hyperelliptic involution $\theta: M \longrightarrow M$, and if $\Theta: M \times M \longrightarrow M \times M$ is the complex analytic automorphism of period two defined by $\Theta(p,q) = (\theta q, \theta p)$, then the mapping $\Psi: M \times M \longrightarrow W_1 - W_1$ maps the diagonal $D \subset M \times M$ to zero and otherwise exhibits $(M \times M)\setminus D$ as a two sheeted branched analytic covering of the variety $(W_1 - W_1)\setminus 0$, with covering translation $\Theta: M \times M \longrightarrow M \times M$ and with branch locus $B = \{(p,\theta p) \subset M \times M\}$. Note that the branch locus B and the diagonal D are one dimensional complex analytic submanifolds of $M \times M$, each being analytically homeomorphic to the surface M ;

and that these submanifolds meet precisely at the Weierstrass points on M. The differential of the mapping $\Psi: M \times M \longrightarrow J(M)$ at a point $(p,q) \in M \times M$ is evidently $d\Psi_{(p,q)} = (d\psi_p, -d\psi_q)$; the mapping Ψ is consequently singular at the point (p,q) , that is to say, $\text{rank } d\Psi_{(p,q)} < 2$, if and only if either $p = q$ or $p \neq q$ and $\gamma(\kappa\zeta_p^{-1}\zeta_q^{-1}) > g-2$. In the latter case it follows from the Riemann-Roch theorem that $\gamma(\zeta_p\zeta_q) > 1$, hence that $\zeta_p\zeta_q \in W_2^2$; the Riemann surface M is consequently hyperelliptic, and $q = \theta p$ for the hyperelliptic involution θ . Thus the mapping Ψ is singular only at points of the diagonal, or in the hyperelliptic case, at the branch points of the mapping Ψ as well as points of the diagonal; and that is of course the best that could be expected, although something further can be added. If the surface M is hyperelliptic and $(p,\theta p) \in B$ is a branch point of the mapping $\Psi: M \times M \longrightarrow W_1 - W_1$, then introducing a coordinate neighborhood U of the point p in M with local coordinate z , the product neighborhood $U \times U$ can be used to coordinatize an open neighborhood of $(p,\theta p)$ in the manifold $M \times M$ by the mapping $(z_1 z_2) \in U \times U \longrightarrow (z_1, \theta z_2) \in U \times \theta U$; and in terms of these local coordinates (z_1, z_2) , the automorphism $\Theta: U \times \theta U \longrightarrow U \times \theta U$ evidently has the form $\Theta(z_1,z_2) = (z_2,z_1)$, so that by Theorem 9 the quotient space $(M \times M)/\Theta$ has the natural structure of two dimensional complex analytic manifold. It is fairly easy to verify that the induced complex analytic mapping $\Psi: (M \times M)/\Theta \longrightarrow J(M)$ remains nonsingular at the branch points in B outside of the diagonal D ; the details will be left to the reader. The subvariety $W_1 - W_1 \subseteq J(M)$ is thus nonsingular except perhaps at the origin; and if $g > 2$ the origin is definitely a singular point, indeed a singular point of imbedding dimension g , as an immediate consequence of the argument in the proof of Theorem 11.

Similar results can be established for the mappings $\Psi\colon M^{(r)} \longrightarrow J(M)$ for $r > 1$ as well; but these results are somewhat more complicated to state, and perhaps of somewhat less interest, so that details will be left to the reader. Let it suffice merely to observe that if $\Psi(\mathscr{D}_1, \mathscr{D}_1') = \Psi(\mathscr{D}_2, \mathscr{D}_2')$ for divisors $\mathscr{D}_i, \mathscr{D}_i'$ in $M^{(r)}$, then $\mathscr{D}_1 + \mathscr{D}_2' \approx \mathscr{D}_1' + \mathscr{D}_2$; if these latter two divisors coincide and the points $(\mathscr{D}_1, \mathscr{D}_1')$ and $(\mathscr{D}_2, \mathscr{D}_2')$ are distinct then the divisors $\mathscr{D}_1$ and $\mathscr{D}_1'$ must have some points in common, while if the two divisors are distinct but linearly equivalent they are the divisors of two separate sections of a line bundle in W_{2r}^2 . Thus for $r > 1$, in place of the diagonal $D \subset M \times M$ there occurs the irreducible analytic subvariety $\tilde{D} \subset M^{(r)} \times M^{(r)}$ of dimension $2r-1$ consisting of pairs of divisors $(\mathscr{D}, \mathscr{D}')$ having at least one point in common, while in place of hyperelliptic surfaces as exceptional cases there occur surfaces which can be represented as branched coverings of $\mathbb{P}^1$ having at most $2r$ sheets.

(c) There are of course other invariants than the imbedding dimension which can be associated to the singularities of complex analytic varieties; but the only other invariant that seems to have been examined in detail in the present context is the multiplicity, the consideration of which was introduced by Riemann. In general, if x is any point on a hypersurface V (that is, on a complex analytic subvariety V of codimension one) in a complex manifold W, then the ideal of all germs of analytic functions on W at the point x which vanish on V can be generated by a single analytic function f; and if all partial derivatives of that function f at the point x with respect to coordinates in W are zero, up to and including order $\nu-1$, then the hypersurface V is said to have multiplicity at least ν at the point x, written $\text{multiplicity}_x(V) \geqq \nu$. Since the function f is unique up to multiplication by a nonvanishing analytic function, it is apparent that the multiplicity is well defined, in the sense of being independent of the choice of the function f. The hypersurface is said to have multiplicity ν if it has multiplicity at least ν but not at least $\nu+1$. It must be emphasized that this definition of multiplicity is only given for hypersurfaces; the extension to subvarieties of lower dimension is somewhat more complicated.

It is quite easy to see that <u>the hypersurface</u> $W_{g-1} \subset J(M)$ <u>has multiplicity at least</u> ν <u>at each point of the subset</u> W^{ν}_{g-1}, that is, that $\text{multiplicity}_x(W_{g-1}) \geqq \nu$ whenever $x \in W^{\nu}_{g-1}$. To demonstrate this, select any analytic function f in an open neighborhood of the point x in $J(M)$ such that f generates the ideal of the hypersurface W_{g-1} at x. Recall from Lemma 3 that

$W^{\nu}_{g-1} = W_{g-1} \ominus (W_{\nu-1} - W_{\nu-1})$; hence if $x \in W^{\nu}_{g-1}$ then necessarily $x + W_{\nu-1} - W_{\nu-1} \subseteq W_{g-1}$. Fixing any points $p_i \in M$ and letting z_i be local coordinates near the points p_i on the Riemann surface M, note that $f(x+\varphi(z_1 +\dots+ z_{\nu-1}) - \varphi(p_1 +\dots+ p_{\nu-1}) = 0$ since $x+\varphi(z_1 +\dots+ z_{\nu-1}) - \varphi(p_1 +\dots+ p_{\nu-1}) \in x+W_{\nu-1} - W_{\nu-1} \subseteq W_{g-1}$; then differentiating this identity with respect to the variables z_i at the points p_i , it follows from the chain rule that

$$(17) \qquad \sum_{j_1,\dots j_{\nu-1}=1}^{g} \frac{\partial^{\nu-1} f(x)}{\partial w_{j_1} \dots \partial w_{j_{\nu-1}}} \cdot w'_{j_1}(p_1) \dots w'_{j_{\nu-1}}(p_{\nu-1}) = 0$$

for all points $p_i \in M$, where $(w_1,\dots,w_g)$ are the usual standard local coordinates on $J(M)$. The vectors $(w'_1(p),\dots,w'_g(p))$ for various points $p \in M$ span the full space $\mathbb{C}^g$, since the Abelian differentials on M are linearly independent; so it follows immediately from (17) that

$$\frac{\partial^{\nu-1} f(x)}{\partial w_{j_1} \dots \partial w_{j_{\nu-1}}} = 0 \quad \text{for arbitrary} \quad j_1,\dots,j_{\nu-1} ,$$

hence that $\text{multiplicity}_x(W_{g-1}) \geqq \nu$ as desired.

The converse assertion is also true, namely that if the hypersurface $W_{g-1} \subset J(M)$ has multiplicity at least ν at a point x then $x \in W^{\nu}_{g-1}$; and therefore W^{ν}_{g-1} is the set of those points at which the hypersurface has multiplicity precisely ν . The proof of the converse assertion is rather more difficult; a sketch was given by Riemann, and more detailed proofs can be found in [6], [8], and [16].

(d) For a hyperelliptic Riemann surface note that $\mathcal{S}(W_{g-1}) = W^2_{g-1} = W_{g-3} - e$; and iterating this observation, $\mathcal{S}(\mathcal{S}(W_{g-1})) = W^2_{g-3} - e = W_{g-5} - 2e$, and so on. If g is an even integer, say $g = 2n$, then eventually

$$\mathcal{S}(\mathcal{S}(\ldots \mathcal{S}(W_{g-1})\ldots)) = W_1 - (n-1)\cdot e ;$$

thus the subvariety W_1 , which is analytically homeomorphic to the Riemann surface, is determined uniquely up to translation by the subvariety W_{g-1} itself. If g is an odd integer, say $g = 2n+1$, then eventually

$$\mathcal{S}(\mathcal{S}(\ldots \mathcal{S}(W_{g-1})\ldots)) = - n\cdot e ,$$

so that a multiple of the hyperelliptic point is determined by the subvariety W_{g-1} itself.

(e) It is a familiar result in the theory of functions of several complex variables that a complex analytic subvariety V of an r-dimensional complex manifold that is defined by the vanishing of n analytic functions has dimension at least $r-n$; indeed each irreducible component of V must separately have dimension at least $r-n$.

(f) It is possible to describe the subvarieties $W^{\nu}_r \subseteq J(M)$ in a manner quite parallel to the description of the subvarieties $G^{\nu}_r \subseteq M^{(r)}$ provided by equation (7), and thereby to obtain a more direct proof of Theorem 14(b) patterned upon the proof of Theorem 14(a) and also an alternative approach to the topological analysis of the condition that the subvarieties W_r^ν be empty sketched briefly at the end of section (e) of this chapter.

For any line bundle $\xi \in P(M)$ and any fixed index r it follows from the Riemann-Roch theorem that

$$(18) \qquad \gamma(\xi\zeta_{p_o}^r) = \gamma(\kappa\xi^{-1}\zeta_{p_o}^{-r}) + r+1-g .$$

Now taking a fixed divisor $\mathfrak{d} = p_1 + \ldots + p_n \in M^{(n)}$ formed from n distinct points of the Riemann surface M, it is quite evident that

$$(19) \qquad \gamma(\kappa\xi^{-1}\zeta_{p_o}^{-r}) = \dim\{f \in \Gamma(M,\mathcal{O}(\kappa\xi^{-1}\zeta_{p_o}^{-r}\zeta_{\mathfrak{d}}))\,|\,\mathfrak{d}(f) \geqq \mathfrak{d}\} .$$

If $r < n$ then $c(\kappa\xi^{-1}\zeta_{p_o}^{-r}\zeta_{\mathfrak{d}}) = 2g-2-r+n > 2g-1$ so that $\gamma(\kappa\xi^{-1}\zeta_{p_o}^{-r}\zeta_{\mathfrak{d}}) = g-1+n-r$, which is of course independent of the choice of the line bundle $\xi \in P(M)$. Choosing for each $\xi \in P(M)$ a basis $\{f^i(\xi)\}$, $1 \leq i \leq g-1+n-r$, for the vector space $\Gamma(M,\mathcal{O}(\kappa\xi^{-1}\zeta_{p_o}^{-r}\zeta_{\mathfrak{d}}))$, it follows easily from (19) that

$$(20) \qquad \gamma(\kappa\xi\zeta_{p_o}^{-r}) = g-1+n-r - \operatorname{rank}\{f^i(\xi,p_j)\} .$$

In this formula $\{f^i(\xi,p_j)\}$ is the matrix of $g-1+n-r$ rows and n columns in which column j consists of the values of the sections $f^i(\xi)$, $1 \leq i \leq g-1+n-r$, at the point p_j when these sections are all represented by complex analytic functions in an open neighborhood of p_j by some local trivialization of the line bundle $\kappa\xi^{-1}\zeta_{p_o}^{-r}\zeta_{\mathfrak{d}}$ over that neighborhood; the rank of that matrix is evidently independent of the choice of local trivialization. Combining equations (18) and (20), it follows that

$$\gamma(\xi\zeta_{p_o}^{r}) = n - \operatorname{rank}\{f^i(\xi,p_j)\} , \tag{21}$$

and therefore that

$$W_r^{\nu} = \{\xi \in P(M) \,|\, \operatorname{rank}\{f^i(\xi,p_j)\} \leqq n-\nu\} . \tag{22}$$

This last result is quite parallel to the description of the subvarieties $G_r^{\nu} \subseteq M^{(r)}$ provided by equation (7), except that as yet nothing has been said about the analyticity of the matrix $\{f^i(\xi,p_j)\}$ as a function of the point $\xi \in P(M)$; but that can be taken care of quite simply by an application of Theorem 16(a). It follows immediately from that theorem that for a sufficiently small open neighborhood V of any point of the manifold $J(M)$ there exist an open covering $\mathcal{U} = \{U_\alpha\}$ of the Riemann surface M and complex analytic mappings $f_\alpha^i\colon V \times U_\alpha \longrightarrow \mathbf{C}$ such that for each fixed $\xi \in V$ the cochains $\{f_\alpha^i(\xi)\} \in C^0(\mathcal{U}, \mathcal{O})$ represent a basis for the space of holomorphic sections of the line bundle $\kappa\xi^{-1}\zeta_{p_o}^{-r}\zeta_{\vartheta}$, when that line bundle is represented in turn by the appropriate cocycle in $Z^1(\mathcal{U}, \mathcal{O}^*)$. Choosing for each point p_j of the divisor ϑ an open set $U_{\alpha_j} \in \mathcal{U}$ containing p_j , the matrix $\{f_{\alpha_j}^i(\xi,p_j)\}$ is a matrix of holomorphic functions of the variable $\xi \in V$; and upon using this matrix in equation (22), the parallel with the earlier discussion of the subvarieties G_r^{ν} is quite evident.

Considering the description (22) in an open neighborhood V of a point $\xi_o \in W_r^{\nu} \setminus W_r^{\nu+1}$, where the matrix $\{f^i(\xi, p_j)\}$ consists of analytic functions of ξ , it follows from Lemma 6 exactly as in the proof of Theorem 14(a) that near ξ_o the subvariety W_r^{ν} can be described as the set of common zeros of $(g-1+n-r-n+\nu)(n-n+\nu) = (g-1-r+\nu)\nu$ analytic functions; hence $\dim_{\xi}(W_r^{\nu}) \geqq g - (g-1-r+\nu)\nu = r\nu - (\nu-1)(g+\nu)$, providing another proof of Theorem 14(b).

(g) A few words should also be added to describe an interpretation of the matrices $\{f^i(\xi, p_j)\}$ appearing in equations (21) and (22), paralleling the interpretation of the differential of the Jacobi mapping $\psi\colon M^{(r)} \longrightarrow J(M)$ as an analytic bundle mapping, as described at the end of section (e) of this chapter; this involves global properties of these matrices as ξ ranges over the full manifold $P(M)$. Introducing open coverings $\{V_A\}$ of the manifold $P(M)$ and $\mathcal{U} = \{U_\alpha\}$ of the Riemann surface M as before, there are analytic mappings $x_{A\alpha\beta}\colon V_A \longrightarrow \mathbf{C}^*$ and $\zeta_{\alpha\beta}\colon U_\alpha \cap U_\beta \longrightarrow \mathbf{C}^*$ such that $x_A(\xi) = \{x_{A\alpha\beta}(\xi)\} \in Z^1(\mathcal{U}, \mathbf{C}^*)$ describes a flat line bundle representing the analytic line bundle ξ^{-1} over M for any fixed $\xi \in V_A$, and that $\{\zeta_{\alpha\beta}\} \in Z^1(\mathcal{U}, \mathcal{O}^*)$ represents the analytic line bundle $\kappa\zeta_{p_o}^{-r}\zeta_{\vartheta}$ over M ; and there are analytic mappings $f^i_{A\alpha}\colon V_A \times U_\alpha \longrightarrow \mathbf{C}$ such that $\{f^i_{A\alpha}(\xi)\} \in \Gamma(M, \mathcal{O}(\kappa\zeta_{p_o}^{-r}\zeta_{\vartheta}))$ represent a basis for the space of holomorphic sections of the line bundle $\kappa\xi^{-1}\zeta_{p_o}^{-r}\zeta_{\vartheta}$ for any fixed $\xi \in V_A$. There are also complex analytic mappings $w_{i\alpha}\colon U_\alpha \longrightarrow \mathbf{C}$

and $s_{ABi}: V_A \cap V_B \longrightarrow \mathbf{C}$ such that $dw_{i\alpha} = \omega_i$ in U_α and that equation (12) holds. It is easy to see that for any fixed point $p \in U_\alpha$ the functions $s_{AB\alpha}(\xi,p) = \exp\{-2\pi i \sum_{i=1}^{g} s_{ABi}(\xi)w_{i\alpha}(p)\}$ form a cocycle $s_\alpha(p) = \{s_{AB\alpha}(p)\} \in Z^1(\{V_A\}, \mathcal{O}^*)$ and hence determine an analytic line bundle over the manifold $P(M)$; and it follows immediately from (13) that the cocycles $s_\alpha(p)$ and $s_\beta(p)$ are cohomologous whenever $p \in U_\alpha \cap U_\beta$, so determine the same line bundle. This line bundle can thus be noted unambiguously by $\Sigma\,(p) \in H^1(J(M), \mathcal{O}^*)$. Note that for each fixed i a constant can be added to all of the functions $w_{i\alpha}$ simultaneously without changing the identity (12); so it can be assumed that $w_{i\alpha_0}(p_0) = 0$, where $p_0 \in U_{\alpha_0}$, and hence that $\Sigma\,(p_0) = 1$. The bundles $\Sigma(p)$ are obviously deformed onto one another as p varies over M , so are all topologically equivalent; and since $\Sigma(p_0) = 1$, it follows that all of these bundles $\Sigma(p)$ are topologically trivial. Now for any fixed point $p \in U_\alpha$ the functions $\{f^i_{A\alpha}(\xi,p)\}$ can be considered as defining a bundle mapping

$$\sigma(p): \bigcup_{\xi \in P(M)} \Gamma(M, \mathcal{O}(\kappa\xi^{-1}\zeta_{p_0}^{-1}\zeta_{\vartheta})) \longrightarrow \Sigma(p)$$

over the subset V_A , by setting

$$(23) \qquad \sigma(p)(\xi, t_1^A, \ldots, t_n^A) = \sum_{i=1}^{n} f^i_{A\alpha}(\xi,p)t_i^A$$

in terms of the local coordinates introduced earlier in the fibre bundle $\bigcup_{\xi \in P(M)} \Gamma(M, \mathcal{O}(\kappa\xi^{-1}\zeta_{p_0}^{-1}\zeta_{\vartheta}))$. Note that whenever $\xi \in V_A \cap V_B$

it follows from (14) and (16) that

$$\sum_{i=1}^{n} f^{i}_{A\alpha}(\xi,p)t^{A}_{i} = s_{AB\alpha}(\xi,p)\cdot \sum_{j=1}^{n} f^{j}_{B\alpha}(\xi,p)t^{B}_{j} \ ;$$

consequently the bundle mapping $\sigma(p)$ is well defined over the full manifold $P(M)$. Altogether then, the matrices $\{f^{i}_{A\alpha_j}(\xi,p_j)\}$ can be viewed as defining a complex analytic bundle mapping

$$(24) \qquad \sigma: \bigcup_{\xi \in P(M)} \Gamma(M, \mathcal{O}(\kappa\xi^{-1}\zeta_{p_o}^{-r}\zeta_{\mathscr{D}})) \longrightarrow \bigoplus_{j=1}^{n} \Sigma(p_j)$$

onto the direct sum of the topologically trivial complex analytic line bundles $\Sigma(p_j)$ over the manifold $P(M)$, as the points p_j range over the divisor $\mathscr{D}$; and equation (22) can be rewritten in the form

$$(25) \qquad W^{\nu}_{r} = \{\xi \in P(M) \,|\, \text{rank}\ \sigma_{\xi} \leqq n-\nu\} \ .$$

Again if there are topological obstructions to the existence of a bundle mapping (24) of rank $> n-\nu$ at all points $\xi \in P(M)$, then the subvarieties W^{ν}_{r} are necessarily nonempty.

References. In addition to the references listed at the end of §2, sources for the material discussed in this chapter were the following.

[11] Farkas, Hershel M., Special divisors and analytic subloci of Teichmüller space, Amer. Jour. Math. 88(1966), 881-901.

[12] Kempf, George, The singularities of certain varieties in the Jacobian of a curve. (Ph.D. thesis, Columbia University, 1971.)

[13] -----, Schubert methods with an application to algebraic curves. (Stichting Mathematisch Centrum, Amsterdam ii + 18 pp., 1971.)

[14] Kleiman, Steven L. and Laksov, Dan, On the existence of special divisors, Amer. Jour. Math. 94(1972), 431-436.

[15] Kodaira, K., Some results in the transcendental theory of algebraic varieties, Annals of Math. 59(1954), 86-134.

[16] Lewittes, Joseph, Riemann surfaces and the theta function, Acta Math. 111(1964), 37-61.

[17] Macdonald, I. G., The Poincaré polynomial of a symmetric product, Proc. Cambridge Phil. Soc. 58(1962), 563-568.

[18] -----, Symmetric products of an algebraic curve, Topology 1 (1962), 319-343.

[19] Mattuck, Arthur, Symmetric products and Jacobians, Amer. Jour. Math. 83(1961), 189-206.

[20] -----, Picard bundles, Illinois Jour. Math. 5(1961), 550-564.

[21] -----, On symmetric products of curves, Proc. Amer. Math. Soc. 13(1962), 82-87.

[22] -----, Secant bundles on symmetric products, Amer. Jour. Math. 87(1965), 779-797.

[23] Mattuck, A. and Mayer, A., The Riemann-Roch theorem for algebraic curves, Annali Scuola Norm. Sup. Pisa 17(1963), 223-237.

[24] Meis, T., Die minimale Blätterzahl der Konkretisierung einer kompakten Riemannschen Fläche, Schriften. Math. Inst. Münster (1960).

[25] Schwarzenberger, R. L. E., Jacobians and symmetric products, Illinois Jour. Math. 7(1963), 257-268.

[26] -----, The secant bundle of a projective variety, Proc. London Math. Soc. 14(1964), 369-384.

The importance of symmetric products of Riemann surfaces in studying their Jacobi varieties has long been recognized, and further references to earlier work can be found in the papers listed in this brief bibliography; symmetric products are particularly useful from an algebraic point of view, as is clear from the well known books of A. Weil on these subjects. Much recent activity even in the purely analytic case has been inspired by the work of Weil; Corollary 4 to Theorem 10 and the proof of Theorem 11 are basically taken from [10] for instance, though with some extensions as given in [8]. The treatment used here owes a good deal to the papers and lectures of Martens. Theorem 12, and especially the proof of Lemma 5, were inspired by [4]; and the proof of the extended Clifford theorem in section (d) was taken directly from [4]. The topological analysis mentioned at the end of section (e) can be found carried out in [12], [13], and [14], but is done there in the context of the Jacobi mapping as a fibration as discussed in the notes to section (g); the corresponding treatment for the differential of the Jacobi mapping seems new.

§4. Intersections in Jacobi varieties and Torelli's theorem.

(a) Intersections of subvarieties of positive divisors in the Jacobi variety have frequently been considered in the preceding discussion, as for example in the formulas from Lemmas 1 and 2,

$$W_{s-r} = W_s \ominus W_r = \bigcap_{u \in W_r} (W_s - u) \ , \quad 0 \leqq r \leqq s \leqq g-1 \ ,$$

$$W_r^{\nu} = W_{r-\nu+1} \ominus (-W_{\nu-1}) = \bigcap_{u \in W_{\nu-1}} (W_{r-\nu+1} + u) \ , \quad \nu \leqq r+1 \ .$$

These are infinite intersections, though, and it is of some interest also to examine analogous finite intersections, as indicated by the discussion in §3(c). The proper investigation of these intersections requires some notion of intersection multiplicity, either analytical or topological; but leaving such complications aside for a later treatment, it is still possible to derive a number of interesting and useful results merely involving intersections in the point set sense.

Theorem 17(a). For any index $r < g$ and any points $u, v \in W_1$ with $u \neq v$,

$$(W_r + u) \cap (W_r + v) = W_{r+1}^2 \cup (W_{r-1} + u + v) \ .$$

Proof. Since $u, v \in W_1$ it follows that $W_{r-1} + u \subseteq W_r$ and $W_{r-1} + v \subseteq W_r$, and therefore that $W_{r-1} + u + v \subseteq (W_r + v) \cap (W_r + u)$; and it follows from Lemma 2 that $W_{r+1}^2 = W_r \ominus (-W_1) = \bigcap_{x \in W_1} (W_r + x) \subseteq$ $\subseteq (W_r + u) \cap (W_r + v)$. There to complete the proof it is only necessary to show that

$$(W_r + u) \cap (W_r + v) \subseteq W^2_{r+1} \cup (W_{r-1} + u+v) .$$

Write $u = \varphi(p)$, $v = \varphi(q)$ for some points $p,q \in M$ with $p \neq q$, and note that any point $x \in (W_r + u) \cap (W_r + v)$ can be written

$$x = \varphi(p_1 + \ldots + p_r + p) = \varphi(q_1 + \ldots + q_r + q)$$

for some points $p_i, q_i \in M$; it follows from Abel's theorem that these two divisors are linearly equivalent, that is, that

$$p_1 + \ldots + p_r + p \approx q_1 + \ldots + q_r + q .$$

If these two divisors are actually identical, then since $p \neq q$ it follows that q must coincide with one of the points p_i, say $q = p_r$; and then

$$x = \varphi(p_1 + \ldots + p_{r-1} + q+p) \in W_{r-1} + v+u .$$

If these two divisors are distinct, then necessarily

$$\gamma(\zeta_{p_1} \cdots \zeta_{p_r}\zeta_p) \geqq 2 ;$$

and then

$$x = \varphi(p_1 + \ldots + p_r + p) \in W^2_{r+1} .$$

Thus $x \in (W_{r-1} + u+v) \cup (W^2_{r+1})$, and the proof is thereby concluded.

Recall from Theorem 7 that $\dim W^2_{r+1} \leqq r-1$ for $1 \leqq r \leqq g-2$, and from Theorem 13 that if $\dim W^2_{r+1} = r-1$ for any index r in the interval $1 \leqq r \leqq g-3$ then the surface is necessarily hyperelliptic. Hence an immediate consequence of Theorem 17(a) is that on a Riemann surface that is not hyperelliptic the analytic subvariety W^2_{r+1} can be characterized as the union of those irreducible components of the intersection $(W_r + u) \cap (W_r + v)$ having dimension strictly less than $r-1$, for any two distinct points $u,v \in W_1$. On the other hand for a hyperelliptic surface $W^2_{r+1} = W_{r-1} - e$, where e is the hyperelliptic point; and in this case

$$(W_r + u) \cap (W_r + v) = (W_{r-1} + u+v) \cup (W_{r-1} - e) .$$

Of course in general

$$W^2_{r+1} = \bigcap_{u \in W_1} (W_r + u) ,$$

with an infinite intersection.

In the special case that $r = g-1$ the intersection in Theorem 17(a) is always purely $(g-2)$-dimensional. Indeed recalling from the Riemann-Roch theorem in the form of equation (13) of §2 that $W^2_g = k - W_{g-2}$ where k is the canonical point, this case of the theorem can evidently be rewritten as follows.

Theorem 17(b). For any points $u,v \in W_1$ with $u \neq v$,

$$W_{g-1} \cap (W_{g-1} + u-v) = (W_{g-2} + u) \cup (k-v - W_{g-2}) .$$

(b) Next consider an intersection of the form $W_{g-1} \cap (W_{g-1} + u)$ for an arbitrary nonzero point $u \in J(M)$. It is a familiar result from the theory of functions of several complex variables that each irreducible component of this intersection is of dimension $g-2$;

and it is of some interest to compare this general intersection with the special case described in Theorem 17(b), in which $u \in W_1 - W_1$.

Introducing the Jacobi mapping $\varphi: M^{g-1} \longrightarrow W_{g-1}$, the inverse image $\varphi^{-1}(W_{g-1} \cap (W_{g-1} + u)) \subset M^{g-1}$ is an analytic subvariety each irreducible component of which is also of dimension $g-2$. In view of this it is quite natural to attempt to describe this latter subvariety in terms of its projection to the first factor in the product representation $M^{g-1} = M^{g-2} \times M$; indeed the reason for introducing the product M^{g-1} in place of the variety W_{g-1} is precisely the possibility of using this sort of factorization of the product. Actually, though, it is even more convenient to consider in place of the Jacobi mapping the analytic mapping $\theta: M^{(g-2)} \times M \longrightarrow W_{g-1}$ resulting from the natural factorization of the Jacobi mapping

$$M^{g-1} \longrightarrow M^{(g-2)} \times M \longrightarrow M^{(g-1)} \xrightarrow{\psi} W_{g-1} ;$$

thus

$$\theta(p_1 + \dots + p_{g-2}, p) = \varphi(p_1 + \dots + p_{g-2}) + \varphi(p) = \varphi(p_1 + \dots + p_{g-2} + p) .$$

In terms of this mapping set

$$V = \theta^{-1}(W_{g-1} \cap (W_{g-1} + u)) \subset M^{(g-2)} \times M ,$$

and let $\pi: V \longrightarrow M^{(g-2)}$ be the analytic mapping induced by the natural projection $M^{(g-2)} \times M \longrightarrow M^{(g-2)}$. Note that the image under θ of each irreducible component of V is an irreducible component of the intersection $W_{g-1} \cap (W_{g-1} + u)$; but of course several irreducible components of V may have the same image under

θ , since this mapping is not one-to-one. The relevant properties of this subvariety V will be described in the following sequence of lemmas.

Lemma 7(a). For a divisor $\mathfrak{d} \in M^{(g-2)}$ such that $\varphi(\mathfrak{d}) \in W_{g-2} \cap (W_{g-2}+u)$, then $\mathfrak{d} \times M \subseteq V \subset M^{(g-2)} \times M$; while for any other divisor $\mathfrak{d} \in M^{(g-2)}$ there are only g points $p_i \in M$ (not necessarily distinct) such that $(\mathfrak{d}, p_i) \in V \subset M^{(g-2)} \times M$, and they are uniquely characterized by the condition that $\varphi(\mathfrak{d} + p_1 + \ldots + p_g) = k+u$.

Proof. For any divisor $\mathfrak{d} \in M^{(g-2)}$ and any point $p \in M$ note that $(\mathfrak{d}, p) \in V \subset M^{(g-2)} \times M$ if and only if $\theta(\mathfrak{d}, p) = \varphi(\mathfrak{d} + p) \in W_{g-1} + u$, hence if and only if $\varphi(\mathfrak{d} + p) - \eta \in W_{g-1}$; and letting $\eta \in P(M)$ be the line bundle corresponding to the point $u \in J(M)$, this latter condition is evidently equivalent to the condition that $\gamma(\zeta_{\mathfrak{d}} \zeta_p \eta^{-1}) \geqq 1$, or since $\gamma(\zeta_{\mathfrak{d}} \zeta_p \eta^{-1}) = \gamma(\kappa\eta\zeta_{\mathfrak{d}}^{-1}\zeta_p^{-1})$ by the Riemann-Roch theorem, to the condition that $\gamma(\kappa\eta\zeta_{\mathfrak{d}}^{-1}\zeta_p^{-1}) \geq 1$. Note that $\gamma(\kappa\eta\zeta^{-1}) \geqq 1$ since $c(\kappa\eta\zeta_{\mathfrak{d}}^{-1}) = g$. If $\gamma(\kappa\eta\zeta_{\mathfrak{d}}^{-1}) = 1$ there is a unique holomorphic section of the line bundle $\kappa\eta\zeta_{\mathfrak{d}}^{-1}$, and $\gamma(\kappa\eta\zeta_{\mathfrak{d}}^{-1}\zeta_p^{-1}) \geqq 1$ precisely when p is one of the zeros of this section; and writing the divisor of this section out as $p_1 + \ldots + p_g$ it of course follows that $\kappa\eta\zeta_{\mathfrak{d}}^{-1} = \zeta_{p_1} \cdots \zeta_{p_g}$, or equivalently that $\varphi(\mathfrak{d} + p_1 + \ldots + p_g) = k+u$. On the other hand if $\gamma(\kappa\eta\zeta_{\mathfrak{d}}^{-1}) \geqq 2$ then $\gamma(\kappa\eta\zeta_{\mathfrak{d}}^{-1}\zeta_p^{-1}) \geqq 1$ for all points $p \in M$; thus in this case

$(\mathfrak{d}, p) \in V$ for all points $p \in M$. Note that $\gamma(\kappa\eta\zeta_{\mathfrak{d}}^{-1}) \geqq 2$ precisely when $k+u - \varphi(\mathfrak{d}) \in W_g^2 = k - W_{g-2}$, hence precisely when $\varphi(\mathfrak{d}) \in W_{g-2} \cap (W_{g-2} + u)$; that suffices to conclude the proof of the lemma.

Letting V_i be an irreducible component of the analytic subvariety $V \subset M^{(g-2)} \times M$, set $X_i = \{\mathfrak{d} \in M^{(g-2)} | \mathfrak{d} \times M \subseteq V_i\}$. Note that X_i is evidently a complex analytic subvariety of the manifold $M^{(g-2)}$; for there exist an open neighborhood U' of any given divisor $\mathfrak{d}_o \in M^{(g-2)}$ and an open covering $\{U''_\nu\}$ of the Riemann surface M so that in each set $U' \times U''_\nu \subseteq M^{(g-2)} \times M$ the subvariety $V_i \cap (U' \times U''_\nu)$ is defined by the vanishing of some holomorphic function $f_\nu(\mathfrak{d}, p)$, and $X_i \cap U' = \{\mathfrak{d} \in U' | f_\nu(\mathfrak{d}, p) = 0$ for all $p \in U''_\nu$ and all $\nu\}$. If $\dim X_i = g-3$ then X_i is necessarily an irreducible analytic subvariety of $M^{(g-2)}$ and $V_i = X_i \times M$, since $X_i \times M$ is an analytic subvariety of dimension $g-2$ contained within the irreducible subvariety V_i which is also of dimension $g-2$. In this case V_i will be called an <u>exceptional component</u> of the analytic subvariety V. Note then that $Y_i = \varphi(X_i)$ is an irreducible component of the intersection $W_{g-2} \cap (W_{g-2} + u)$ of dimension $g-3$, recalling Lemma 7(a), and that $\theta(V_i) = \theta(X_i \times M) = Y_i + W_1$ is an irreducible component of the intersection $W_{g-1} \cap (W_{g-1} + u)$. If $\dim X_i < g-3$ then the mapping $\pi\colon V_i \longrightarrow M^{(g-2)}$ has as image the entire manifold $M^{(g-2)}$; for the image $\pi(V_i)$ is always an analytic subvariety of $M^{(g-2)}$ as a consequence of the proper mapping theorem, and is clearly a proper analytic subvariety only in the preceding case. Indeed the restriction

$$\pi: V_i \setminus (V_i \cap \pi^{-1}(X_i)) \longrightarrow M^{(g-2)} \setminus X_i$$

is a branched analytic covering of at most g sheets; for the inverse image $\pi^{-1}(\mathfrak{d}) \cap V_i$ is a proper analytic subvariety of $\mathfrak{d} \times M$ and hence a finite number of points whenever $\mathfrak{d} \in M^{(g-2)} \setminus X_i$, so that the mapping π is a branched analytic covering, and in general $\pi^{-1}(\mathfrak{d}) \cap V_i$ contains at most g points as a consequence of Lemma 7(a). Note particularly that the exceptional set X_i has codimension at least 2 in the manifold $M^{(g-2)}$. In this case V_i will be called a <u>standard component</u> of the analytic subvariety V. If $\dim W_{g-2} \cap (W_{g-2} + \dot{u}) < g-3$ then all the irreducible components of V are necessarily standard components; and this would be expected to be the usual situation, of course. The exceptional components are exceptional in another sense as well, as follows.

<u>Lemma 7(b).</u> For any exceptional component V_i of the analytic subvariety $V \subset M^{(g-2)} \times M$ there exists a standard component V_j of that subvariety such that $\theta(V_j) = \theta(V_i)$.

Proof. As a preliminary observation it will be demonstrated that for any index $s \leqq g-1$, any irreducible analytic subvariety $X \subset W_s$ of dimension $s-1$, and any divisor $\mathfrak{d} \in M^{(s-1)}$, there exists a point $p \in M$ such that $\varphi(\mathfrak{d} + p) \in X$. Note that the desired lemma follows immediately from the special case $s = g-1$ of this observation. For if V_i is any exceptional component of

V then applying this observation to the irreducible analytic subvariety $X = \theta(V_i) \subset W_{g-1}$ it is evident that the image of the natural projection of the analytic subvariety $\theta^{-1}(X) \subset M^{(g-2)} \times M$ to the first factor is all of $M^{(g-2)}$; since $\theta^{-1}(X)$ is a union of some irreducible components of V clearly one of these components must be a standard component V_j, and then of course $\theta(V_j) = \theta(V_i)$ as desired.

The observation itself will be demonstrated by induction on the index s, the case $s = 1$ being quite trivial. Introducing for any index s the analytic mapping $\theta_s: M^{(s-1)} \times M \longrightarrow W_s$ defined by $\theta_s(\mathfrak{I}, p) = \varphi(\mathfrak{I} + p)$, the subset $\theta_s^{-1}(X) \subset M^{(s-1)} \times M$ is an irreducible analytic subvariety each component of which is of dimension $s-1$; and if $\tilde{X} \subset M^{(s-1)} \times M$ is any one of these irreducible components then $\theta_s(\tilde{X}) = X$. If the image of the natural projection of the subvariety $\tilde{X} \subset M^{(s-1)} \times M$ to the first factor is all of $M^{(s-1)}$ the desired result follows immediately. Otherwise the image of that projection is an irreducible complex analytic subvariety $\tilde{Y} \subset M^{(s-1)}$, as a consequence of the proper mapping theorem; and since $\tilde{X} \subseteq \tilde{Y} \times M$ and both these subvarieties are irreducible, necessarily $\tilde{Y}$ is of dimension $s-2$ and $\tilde{X} = \tilde{Y} \times M$. Now the image $Y = \varphi(\tilde{Y}) \subset W_{s-1}$ is an irreducible analytic subvariety of dimension $s-2$ to which the induction step can be applied; so for any divisor $\mathfrak{I} = \mathfrak{I}' + p_{s-1} \in M^{(s-1)}$, where $\mathfrak{I}' \in M^{(s-2)}$, there exists a point $p \in M$ such that $\varphi(\mathfrak{I}' + p) \in Y$. Now

$X = \theta_s(\tilde{X}) = \theta_s(\tilde{Y}\times M) = Y + W_1$, and hence $\varphi(\vartheta + p) = \varphi(\vartheta' + p + p_{s-1}) =$ $= \varphi(\vartheta' + p) + \varphi(p_{s-1}) \in Y + W_1 \subseteq X$; that suffices to conclude the proof of this observation, and thereby also the proof of the lemma.

Considering now a standard component V_i of the analytic subvariety $V \subset M^{(g-2)} \times M$, as already noted the restricted projection mapping

$$\pi\colon V_i \setminus (V_i \cap \pi^{-1}(X_i)) \longrightarrow M^{(g-2)} \setminus X_i$$

is a branched analytic covering of ν_i sheets for some index $\nu_i \leqq g$ and some analytic subvariety $X_i \subset M^{(g-2)}$ of codimension at least 2. For any divisor $\vartheta \in M^{(g-2)} \setminus X_i$ there are ν_i points $(\vartheta, p_j(\vartheta)) \in V_i \subset M^{(g-2)} \times M$, $(j = 1,\dots,\nu_i)$; these points are generally distinct, although there may be coincidences when the divisor ϑ lies in a proper analytic subvariety of $M^{(g-2)} \setminus X_i$. It is a familiar result from the theory of functions of several complex variables that for any analytic function f in an open subset of $M^{(g-2)} \times M$ the symmetric expression $f(p_1(\vartheta)) + \dots + f(p_{\nu_i}(\vartheta))$ when viewed as a function of the divisor $\vartheta \in M^{(g-2)}$ is holomorphic wherever it is well defined. An immediate consequence of this is that the mapping $\theta_i\colon M^{(g-2)} \setminus X_i \longrightarrow J(M)$ defined by

$$(1) \quad \theta_i(\vartheta) = \varphi(p_1(\vartheta)) + \dots + \varphi(p_{\nu_i}(\vartheta)) = \varphi(p_1(\vartheta) + \dots + p_{\nu_i}(\vartheta))$$

is a complex analytic mapping; and since the subvariety X_i is of codimension at least 2 it follows from the extended Riemann removable singularities theorem that the mapping θ_i extends to a complex analytic mapping $\theta_i: M^{(g-2)} \longrightarrow J(M)$. In the notes to §2(a) though it was noted that any such mapping must necessarily be of the form

$$\theta_i(\mathfrak{d}) = C_i\varphi(\mathfrak{d}) + c_i \tag{2}$$

where $C_i: J(M) \longrightarrow J(M)$ is an endomorphism of the Abelian Lie group $J(M)$ (described by a linear transformation $C_i: \mathbb{C}^g \longrightarrow \mathbb{C}^g$ where $J(M) = \mathbb{C}^g/\mathcal{L}$), $c_i \in J(M)$ is a point of the Jacobi variety, and φ is as usual the Jacobi mapping. The mapping θ_i is a convenient tool in this investigation, and the following properties are particularly useful.

<u>Lemma 7(c).</u> If $C_i = 0$ then necessarily $u \in W_{\nu_i} - W_{\nu_i}^{\nu_i}$, and in particular $u \in W_1 - W_1$.

Proof. If $C_i = 0$ then $\varphi(p_1(\mathfrak{d}) + \ldots + p_{\nu_i}(\mathfrak{d})) = \theta_i(\mathfrak{d}) = c_i$ is constant as $\mathfrak{d}$ varies over $M^{(g-2)}\setminus X_i$; and recalling Lemma 7(a), it follows that the divisor $p_1(\mathfrak{d}) + \ldots + p_{\nu_i}(\mathfrak{d}) = \mathfrak{d}_o \in M^{(\nu_i)}$ is also constant as $\mathfrak{d}$ varies over $M^{(g-2)}$ but $\varphi(\mathfrak{d}) \notin W_{g-2} \cap (W_{g-2} + u)$, indeed that for any such divisor $\mathfrak{d} \in M^{(g-2)}$ there will exist a divisor $\tilde{\mathfrak{d}} \in M^{(g-\nu_i)}$ such that $\varphi(\mathfrak{d}_o + \mathfrak{d} + \tilde{\mathfrak{d}}) = k+u$. Now this last condition is equivalent to the condition that $\gamma(\kappa\eta\zeta_{\mathfrak{d}}^{-1}\zeta_{\mathfrak{d}}^{-1}) \geqq 1$ for any such divisor $\mathfrak{d} \in M^{(g-2)}$, and this in turn as in the proof of Lemma 2 to the condition that $\gamma(\kappa\eta\zeta_{\mathfrak{d}_o}^{-1}) \geqq g-1$, or by the Riemann-Roch theorem that $\gamma(\eta^{-1}\zeta_{\mathfrak{d}_o}) \geqq \nu_i$; and this finally merely means that $\varphi(\mathfrak{d}_o) - u \in W_{\nu_i}^{\nu_i}$, hence that $u \in W_{\nu_i} - W_{\nu_i}^{\nu_i}$. This suffices to conclude the proof of the lemma.

If the linear transformation $C_i: \mathbb{C}^g \longrightarrow \mathbb{C}^g$ representing the endomorphism $C_i: J(M) \longrightarrow J(M)$ is of rank r_i then the image $T_i = C_i(J(M)) \subseteq J(M)$ is a connected complex analytic submanifold of dimension r_i in $J(M)$, indeed is a Lie subgroup of the Abelian Lie group $J(M)$, so that it too is a compact complex torus.

Lemma 7(d). The image of the analytic mapping $\theta_i: M^{(g-2)} \longrightarrow J(M)$ is the irreducible analytic subvariety

$$\theta_i(M^{(g-2)}) = C_i(W_{g-2}) + c_i \subseteq T_i + c_i \subseteq J(M)$$

of dimension $d_i = \min(r_i, g-2)$; and if $d_i = r_i$ then $\theta_i(M^{(g-2)}) = T_i + c_i$. Furthermore $d_i \leq \nu_i$; and if equality holds then $d_i = \nu_i = g-2$, $r_i = g$, and $\theta_i(M^{(g-2)}) = W_{g-2}$.

Proof. It follows immediately from (2) that $\theta_i(M^{(g-2)}) = C_i\varphi(M^{(g-2)}) + c_i = C_iW_{g-2} + c_i \subseteq T_i + c_i$; and since these subvarieties are irreducible, they must coincide when their dimensions are equal. The rank of the differential of the analytic mapping $C_i\varphi: M^{(g-2)} \longrightarrow J(M)$ is clearly at most $\min(r_i, g-2)$, and actually attains this maximal value at some divisors $\mathfrak{D} \in M^{(g-2)}$; for otherwise, recalling the form of the differential $d\varphi_{\mathfrak{D}}$, there exists a basis $\{\omega_j = dw_j\}$ for the Abelian differentials on M such that the $r_i \times (g-2)$ matrix $\{w_j'(p_k): j = 1, \ldots, r_i;\ k = 1, \ldots, g-2\}$ is singular for all divisors $\mathfrak{D} = p_1 + \ldots + p_{g-2} \in M^{(g-2)}$ and that is easily seen to be impossible. The image subvariety $\theta_i(M^{(g-2)}) = C_i\varphi(M^{(g-2)}) + c_i$ is consequently of dimension $d_i = \min(r_i, g-2)$, and the first assertion is thereby demonstrated. Next it follows immediately from (1) that $\theta_i(M^{(g-2)}) \subseteq W_{\nu_i}$, hence that $d_i \leqq \nu_i$; and again if these dimensions are equal the two subvarieties must coincide. However if $\theta_i(M^{(g-2)}) = C_i(W_{g-2}) + c_i = W_{\nu_i}$ then the linear transformation C_i must be of rank $r_i = g$, since the subvariety W_{ν_i} cannot be contained in any linear subset of $J(M)$; and it then follows from the first assertion that $\nu_i = d_i = g-2$. That suffices to conclude the proof of the lemma.

Theorem 18. For a Riemann surface of genus $g \geqq 5$ and any point $u \neq 0$ in $J(M)$, the intersection $W_{g-1} \cap (W_{g-1} + u)$ has at most 2 irreducible components; and if it has precisely 2 irreducible components, then either $u \in W_1 - W_1$ and these components are as described in Theorem 17(b), or one of these components is of the form $X + T$ where X is an irreducible subvariety of dimension $g-3$ and T is a Lie subgroup of dimension 1 in the Abelian Lie group $J(M)$.

Proof. Continuing with the notation and terminology introduced in the preceding discussion, let $V_1, \ldots, V_n$ be the standard irreducible components of the analytic subvariety $V \subset M^{(g-2)} \times M$. Assuming then that the intersection $W_{g-1} \cap (W_{g-1} + u)$ has at least 2 irreducible components, it follows immediately from Lemma 7(b) that $n \geqq 2$.

For any divisor $\mathfrak{d} \in M^{(g-2)}$ such that $\varphi(\mathfrak{d}) \notin W_{g-2} \cap (W_{g-2} + u)$ there are g points $p_j \in M$ such that

$(\vartheta, p_j) \in V$ and they are uniquely characterized by the condition that $\varphi(\vartheta + p_1 + \dots + p_g) = k+u$, as in Lemma 7(a). Relabel these points in the form $\{p_j^i\}$, where $(\vartheta, p_1^i), \dots, (\vartheta, p_{\mu_i}^i)$ belong to the irreducible component V_i of the subvariety V. Recall that outside a proper analytic subvariety of $M^{(g-2)}$ the natural projection mapping $V_i \longrightarrow M^{(g-2)}$ is an analytic covering space of ν_i sheets; thus $\mu_i \geqq \nu_i$, and if $\mu_i > \nu_i$ there are necessarily coincidences among the points $\{p_j^i\}$. Indeed since the set of all the divisors $\varphi^{-1}(k+u) \subset M^{(2g-2)}$ is the image of a complex analytic mapping of $\mathbb{P}^{g-2}$ into $M^{(2g-2)}$ by Theorem 10(a), it follows easily by analytic continuation that $\mu_i = \varepsilon_i \nu_i$ for some integer $\varepsilon_i \geqq 1$, and that the points $\{p_j^i\}$ occur in ν_i sets each consisting of ε_i identical points; thus $p_1^i + \dots + p_{\mu_i}^i = \varepsilon_i(p_1^i + \dots + p_{\nu_i}^i)$ and

(3)
$$\sum_{i=1}^{n} \varepsilon_i \nu_i = g .$$

Now with this relabeling note that

$$\begin{aligned} k+u &= \varphi(\vartheta + \sum_{i=1}^{n} \varepsilon_i(p_1^i + \dots + p_{\nu_i}^i)) \\ &= \varphi(\vartheta) + \sum_{i=1}^{n} \varepsilon_i \varphi(p_1^i + \dots + p_{\nu_i}^i) \\ &= \varphi(\vartheta) + \sum_{i=1}^{n} \varepsilon_i \theta_i(\vartheta) \\ &= \varphi(\vartheta) + \sum_{i=1}^{n} \varepsilon_i (C_i \varphi(\vartheta) + c_i) , \end{aligned}$$

using (1) and (2); and since endomorphisms of $J(M)$ are really just suitable linear transformations of the covering space $\mathbb{C}^g$ and hence can be composed as in the algebra of linear transformations, this can be rewritten

$$k+u = (I + \sum_{i=1}^{n} \varepsilon_i C_i)\varphi(\vartheta) + \sum_{i=1}^{n} \varepsilon_i c_i ,$$

where I is the identity transformation. The left hand side is a constant, while the right hand side is a function of ϑ ; and since the images $\varphi(\vartheta)$ can lie in no proper linear subspace of $J(M)$, necessarily

(4) $$I + \sum_{i=1}^{n} \varepsilon_i C_i = 0 ,$$

the trivial linear transformation. Since the identity transformation is thus expressed as a linear combination of the transformations C_i in $\mathbb{C}^g$, it is apparent that the ranks r_i of these transformations must be such that

(5) $$\sum_{i=1}^{n} r_i \geqq g .$$

It is now quite easy to demonstrate that necessarily $r_i \geqq g-2$ for some i ; for if $r_i < g-2$ for all i then it follows immediately from Lemma 7(d) that $d_i = r_i$ and that $d_i < \nu_i$ for all i , and hence from (3) and (5) that

$$g = \Sigma_i \varepsilon_i \nu_i \geqq \Sigma_i \nu_i > \Sigma_i r_i \geqq g ,$$

an evident contradiction. It can therefore be assumed that $r_i \geqq g-2$, indeed from Lemma 7(d) that

(6) $$d_1 = g-2 , \quad r_1 \geqq g-2 , \quad \nu_1 \geqq g-2 .$$

Considering then a second standard irreducible component V_2 of the analytic subvariety $V \subset M^{(g-2)} \times M$, it follows immediately from (3) and (6) that $\nu_2 \leqq 2$. If $\nu_2 = 1$ then from Lemma 7(d) it follows that $d_2 \leqq 1$ and that $d_2 = \nu_2 = 1$ only when $g = 3$, which case has been excluded by hypothesis; furthermore if $0 = d_2 = \min(r_2, g-2)$, then $r_2 = 0$ and from Lemma 7(c) it follows that $u \in W_1 - W_1$, which is one of the desired consequences. If $\nu_2 = 2$ then again from Lemma 7(d) it follows that $d_2 \leqq 2$ and that $d_2 = \nu_2 = 2$ only when $g = 4$, which case has also been excluded by hypothesis; and as before if $d_2 = 0$ then $u \in W_1 - W_1$, which is one of the desired consequences. Therefore the only case left to consider is that in which

$$(7) \qquad d_2 = 1, \quad r_2 = 1, \quad \nu_2 = 2,$$

and in this case $\theta_2(M^{(g-2)}) = T_2 + c_2 \subset J(M)$, where T_2 is a Lie subgroup of the Abelian Lie group $J(M)$ of dimension 1 and c_2 is a point of $J(M)$; note also that $n = 2$, so that V has precisely two irreducible components, and that $r_1 = g$, $\nu_1 = g-2$, and $\varepsilon_1 = \varepsilon_2 = 1$.

To examine this last case more closely, note that for any divisor $\mathfrak{d} = p_1 + \ldots + p_{g-2} \in M^{(g-2)}$ for which $\varphi(\mathfrak{d}) \notin W_{g-2} \cap (W_{g-2} + u)$, the unique divisor in $\varphi^{-1}(k+u) \subset M^{(2g-2)}$ containing $\mathfrak{d}$ can be written as

$$(8) \qquad \mathfrak{d} + (p_1^1 + \ldots + p_{g-2}^1) + (p_1^2 + p_2^2),$$

with the notation introduced above; but this divisor can also be written in the form

(9) $$(p_1 + \dots + p_{g-3} + p_1^2) + (p_1^1 + \dots + p_{g-2}^1) + (p_{g-2} + p_2^2) ,$$

so that as a consequence of Lemma 7(a) the point $(p_1 + \dots + p_{g-3} + p_1^2, p_2^2) \in M^{(g-2)} \times M$ necessarily lies on the subvariety $V \subset M^{(g-2)} \times M$, and therefore

$$\varphi(p_1 + \dots + p_{g-3} + p_1^2 + p_2^2) \in W_{g-1} \cap (W_{g-1} + u) .$$

Now on the other hand from (1), (2), and the preceding observations it follows that

$$\begin{aligned}\varphi(p_1 + \dots + p_{g-3} + p_1^2 + p_2^2) &= \varphi(p_1 + \dots + p_{g-3}) + \theta_2(\mathfrak{d}) \\ &= \varphi(p_1 + \dots + p_{g-3}) + C_2\varphi(\mathfrak{d}) + c_2 \\ &= (I + C_2)\varphi(p_1 + \dots + p_{g-3}) + C_2\varphi(p_{g-2}) + c_2 ,\end{aligned}$$

so that

$$(I + C_2)\varphi(p_1 + \dots + p_{g-3}) + C_2\varphi(p_{g-2}) + c_2 \subseteq W_{g-1} \cap (W_{g-1} + u)$$

for all divisors $\mathfrak{d} \in M^{(g-2)}$; that is to say,

$$(I + C_2)W_{g-3} + T_2 + c_2 \subseteq W_{g-1} \cap (W_{g-1} + u) .$$

Since $I + C_2 = -C_1$ as a consequence of (4) and since C_1 is nonsingular as a consequence of the observation that $r_1 = g$, it follows that $X = (I + C_2)W_{g-3} + c_2$ is an irreducible analytic subvariety of $J(M)$ of dimension $g-3$; and noting that $X \subset X + T_2$, since otherwise $W_{g-3} + (I+C_2)^{-1}T_2 = W_{g-3}$ and this is clearly impossible, it further follows that $X + T_2$ is an irreducible analytic subvariety of $J(M)$ of dimension $g-2$. Therefore $X + T_2$ must be an irreducible component of the intersection $W_{g-1} \cap (W_{g-1}+u)$ of the desired form, and the proof of the theorem is thereby concluded.

There are a number of questions about additional properties of these intersections that come to mind almost at once, even for the case that $g \geqq 5$; but rather than pursuing these matters further here, let it suffice merely to observe that the reducibility of the intersection $W_{g-1} \cap (W_{g-1}+u)$ seems to imply that u is a rather special point of $J(M)$, as indicated in the following.

<u>Corollary 1 to Theorem 18.</u> If for some genus $g \geq 5$ the intersection $W_{g-1} \cap (W_{g-1}+u)$ has 2 irreducible components, then the intersection $W_{g-2} \cap (W_{g-2}+u)$ is of dimension $g-3$.

Proof. If $u \in W_1 - W_1$ then it follows from Theorem 17(a) that one irreducible component of the intersection $W_{g-2} \cap (W_{g-2}+u)$ is a translate of W_{g-3}, and hence is of dimension $g-3$; the only other case in which the intersection $W_{g-1} \cap (W_{g-1}+u)$ has 2 irreducible components is that considered in detail in the last part of the proof of Theorem 18. Continuing with the notation introduced in that proof, for any divisor $\mathfrak{d} \in M^{(g-2)} \setminus E$ where $E = \varphi^{-1}(W_{g-2} \cap (W_{g-2}+u)) \subset M^{(g-2)}$ consider the divisor (8) in $\varphi^{-1}(k+u) \subset M^{(2g-2)}$, and note that it can be rewritten in the form

(10) $\quad (p_1 + \ldots + p_{g-4} + p_1^2 + p_2^2) + (p_1^1 + \ldots + p_{g-2}^1) + (p_{g-3} + p_{g-2})$;

thus setting $\mathcal{S}' = p_1 + \ldots + p_{g-4} + p_1^2 + p_2^2 \in M^{(g-2)}$ it follows from Lemma 7(a) that the points $(\mathcal{S}', p_j^1)$, $(\mathcal{S}', p_{g-3})$, and $(\mathcal{S}', p_{g-2})$ are all contained in the subvariety $V \subset M^{(g-2)} \times M$. If for some divisor $\mathcal{S} \in M^{(g-2)} \backslash E$ the associated divisor $\mathcal{S}'$ also belongs to $M^{(g-2)} \backslash E$ then clearly $F = \{\mathcal{S} \in M^{(g-2)} \backslash E \mid \mathcal{S}' \in E\}$ is a proper analytic subvariety of $M^{(g-2)} \backslash E$; and in this case it is easy to see that the points $(\mathcal{S}', p_j^1)$ are all contained in the irreducible component V_1 of the subvariety V. (For note that for all $\mathcal{S} \in M^{(g-2)} \backslash E$ outside another proper analytic subvariety $G \subset M^{(g-2)} \backslash E$ the points of the divisor (8) are distinct on M, since $\varepsilon_1 = \varepsilon_2 = 1$; then a divisor $\mathcal{S} \in M^{(g-2)} \backslash (E \cup F \cup G)$ can be moved along a closed loop in $M^{(g-2)} \backslash (E \cup F \cup G)$ so that the point $(\mathcal{S}, p_1^1)$ is deformed into any other point $(\mathcal{S}, p_j^1)$ by the natural analytic continuation, and since the same motion can be viewed as deforming the point $(\mathcal{S}', p_1^1)$ into the point $(\mathcal{S}', p_j^1)$ all of these points necessarily belong to the same irreducible standard component of V, indeed to the component V_1 since $g-2 > 2$.) Then from (1) and (2) it follows that

$$\begin{aligned} \varphi(p_1^1 + \ldots + p_{g-2}^1) &= \theta_1(\mathcal{S}) = C_1\varphi(\mathcal{S}) + c_1 \\ &= \theta_1(\mathcal{S}') = C_1\varphi(\mathcal{S}') + c_1 , \end{aligned}$$

hence since C_1 is nonsingular that $\varphi(\mathcal{S}) = \varphi(\mathcal{S}')$ or equivalently that $\varphi(p_{g-3} + p_{g-2}) = \varphi(p_1^2 + p_2^2) = \theta_2(\mathcal{S}) = C_2\varphi(\mathcal{S}) + c_2$;

but this implies that $W_2 \subseteq C_2 T_2 + c_2$, which is clearly impossible. This contradiction then shows that necessarily $\mathfrak{D}' \in E$ for all divisors $\mathfrak{D} \in M^{(g-2)} \backslash E$, or equivalently that

$$\varphi(\mathfrak{D}') = \varphi(p_1 + \ldots + p_{g-4} + p_1^2 + p_2^2)$$

$$= \varphi(p_1 + \ldots + p_{g-4}) + C_2\varphi(\mathfrak{D}) + c_2 \in W_{g-2} \cap (W_{g-2} + u) ;$$

and consequently $W_{g-4} + T_2 + c_2 \subseteq W_{g-2} \cap (W_{g-2} + u)$, so that $W_{g-2} \cap (W_{g-2} + u)$ has dimension $g-3$ as desired. That suffices to conclude the proof of the Corollary.

(c) Finally something should also be said about these intersections for Riemann surfaces of genus $g < 5$. Recall that the hypothesis $g \geq 5$ was used in Theorem 18, in the discussion of the possible values for the various parameters d_i, r_i, ν_i associated to the standard irreducible components V_i of the analytic subvariety $V \subset M^{(g-2)} \times M$, to rule out some exceptional cases; but the presence of these exceptional cases reflects the possibility of the occurrence of some special symmetries in these intersections in the cases $g = 3,4$ and merits some further examination. Referring to the proof of Theorem 18, the exceptional cases were those in which the parameters associated to the second standard component had the values $d_2 = \nu_2 = 1$ in case that $g = 3$ and the values $d_2 = \nu_2 = 2$ in case that $g = 4$. Note also that the discussion of the last case considered in the course of the proof of Theorem 18 must also be modified when $g = 3$, since from $d_2 = 1$ it cannot be concluded that $r_2 = 1$ unless $g-2 > 1$.

<u>Corollary 2 to Theorem 18.</u> For a Riemann surface of genus $g = 3$ and any point $u \neq 0$ in $J(M)$, the intersection $W_{g-1} \cap (W_{g-1} + u)$ has at most 3 irreducible components; if it has precisely 2 irreducible components, then either $u \in W_1 - W_1$ and these components are as described in Theorem 17(b), or one of these components is of the form $T+x$ where $T \subset J(M)$ is a Lie subgroup and x is some point of $J(M)$; and if it has precisely 3 irreducible components, then these all are of the form $T_i + x_i$ where $T_i \subset J(M)$ are linearly independent Lie subgroups and x_i are some points of $J(M)$.

Proof. Referring to the proof of Theorem 18, it is quite easy to verify that when $g = 3$ in addition to the cases in which $u \in W_1 - W_1$ there are only the following three possibilities for the parameters associated to the n standard irreducible components of the analytic subvariety $V \subset M^{(g-2)} \times M$:

(i) $n = 2$ and $d_1 = 1,\ r_1 = 3,\ \nu_1 = 1,\ \varepsilon_1 = 1$,
$d_2 = 1,\ r_2 = 3,\ \nu_2 = 1,\ \varepsilon_2 = 2$;

(ii) $n = 2$ and $d_1 = 1,\ r_1 \geqq 1,\ \nu_1 = 2,\ \varepsilon_1 = 1$,
$d_2 = 1,\ r_2 = 3,\ \nu_2 = 1,\ \varepsilon_2 = 1$;

(iii) $n = 3$ and $d_i = 1,\ r_i = 3,\ \nu_i = 1,\ \varepsilon_i = 1$, $(i = 1,2,3)$.

The last case discussed in the course of the proof of Theorem 18, with the modification as noted above, is subsumed under the exceptional case (ii). It is convenient to discuss these cases separately.

(i) With the notation as in the proof of Theorem 18, it follows from Lemma 7(a) that the divisors in $\varphi^{-1}(k+u) \subset M^{(2g-2)} = M^{(4)}$ can all be written in the form $p_1 + p_1^1 + 2p_1^2$, where $(p_1, p_1^i) \in V_i$ and the points p_1^i are uniquely determined by p_1 whenever $\varphi(p_1) \notin W_1 \cap (W_1 + u)$; and as the point p_1 varies over M, the points p_1^i also vary over all of M for each i, since $d_i = 1$. Thus for any point $p \in M$ there exists a divisor of the form $2p + p' + p''$ in $\varphi^{-1}(k+u) \subset M^{(4)}$. Letting $\eta \in P(M)$ be a line bundle representing $u \in J(M)$ and letting f_1, f_2 be any two linearly independent holomorphic sections of the line bundle $\kappa\eta$, noting that $\gamma(\kappa\eta) = 2$, the divisors in $\varphi^{-1}(k+u) \subset M^{(4)}$ are just the divisors of sections $c_1 f_1 + c_2 f_2$ for arbitrary constants $c_i \in \mathbb{C}$; hence for any point $p \in M$ there are constants $c_i \in \mathbb{C}$ so that $c_1 f_1 + c_2 f_2$ has a double zero at p. Now if these sections are represented by holomorphic functions $f_i(z)$ of a local coordinate z in some coordinate neighborhood on M, then the matrix

$$\begin{pmatrix} f_1(z) & f_1'(z) \\ f_2(z) & f_2'(z) \end{pmatrix}$$

must necessarily be identically singular, where $f_i'(z) = df_i(z)/dz$; but this implies that the functions $f_i(z)$ are linearly dependent, which is impossible. Thus case (i) can never arise after all.

(ii) In this case the divisors in $\varphi^{-1}(k+u) \subset M^{(4)}$ can be written in the form $p_1 + (p_1^1 + p_2^1) + p_1^2$, where $(p_1, p_j^i) \in V_i$ and the points p_j^i are determined uniquely (up to interchange of p_1^1 and p_2^1) by p_1 whenever $\varphi(p_1) \notin W_1 \cap (W_1 + u)$; and again as the point p_1 varies over M so do the points p_j^i, so that for all but finitely many points $p_1 \in M$ it follows that $\varphi(p_j^i) \notin W_1 \cap (W_1 + u)$ for any i,j as well. Then the order of the points in one of these divisors $p_1 + (p_1^1 + p_2^1) + p_1^2$ can be altered by a suitable permutation so that the divisor is rewritten in the form $q_1 + (q_1^1 + q_2^1) + q_1^2$, where again $(q_1, q_j^i) \in V_i$ and $\varphi(q_1) \notin W_1 \cap (W_1 + u)$ but where q_1 is any arbitrarily preassigned one of the four points p_1, p_1^1, p_2^1, or p_1^2 ; and having chosen which one of these four points is to be q_1 the permutation is unique up to an interchange of q_1^1 and q_2^1, so that there are altogether 8 such permutations. It is evident by continuity that the same permutations arise for any divisor in $\varphi^{-1}(k+u) \subset M^{(4)}$, hence that these permutations form a group which will be called the group of admissible permutations for this configuration; this is a subgroup of order 8 and index 3 in the permutation group on 4 letters, is transitive, and contains as a subgroup the cyclic group $\mathbb{Z}_2$ of interchanges of the elements p_1^1 and p_2^1. Actually there is only one possible such subgroup, and it is the subgroup generated by the three permutations of order 2 taking the ordered divisor $p_1 + (p_1^1 + p_2^1) + p_1^2$ to the ordered divisors $p_1 + (p_2^1 + p_1^1) + p_1^2$, $p_1^2 + (p_1^1 + p_2^1) + p_1$, and $p_1^1 + (p_1 + p_1^2) + p_2^1$ respectively; the verification is quite simple and straightforward, if uninteresting, so the details will be left to the reader.

Now applying equations (1) and (2) to the divisors $p_1 + (p_1^1 + p_2^1) + p_1^2$ and $p_1^1 + (p_1 + p_1^2) + p_2^1$ it follows that

$\varphi(p_1^2) = C_2\varphi(p_1) + c_2$ and $\varphi(p_1 + p_1^2) = C_1\varphi(p_1^1) + c_1$, and combining these two equations it further follows that $(I + C_2)\varphi(p_1) + c_2 = C_1\varphi(p_1^1) + c_1$; and since $I + C_1 + C_2 = 0$ as a consequence of (4), this can be rewritten $C_1 \cdot \varphi(p_1 + p_1^1) = c_2 - c_1$. Therefore the irreducible component $\theta(V_1)$ of the intersection $W_{g-1} \cap (W_{g-1} + u)$, which is the image of the analytic continuation of the mapping $p_1 \longrightarrow \varphi(p_1 + p_1^1)$, is contained in a translate of the kernel of the linear transformation C_1 . On the other hand the irreducible component $\theta(V_2)$ is the image of the analytic continuation of the mapping $p_1 \longrightarrow \varphi(p_1 + p_1^2) = (I + C_2)\varphi(p_1) + c_2$, and hence $\theta(V_2) = -C_1(W_1) + c_2$; that component is therefore contained in a translate of the image of the linear transformation C_1 . The rank r_1 of the transformation C_1 is clearly either 1 or 2 though, so that either the kernel of C_1 or the image of C_1 is a linear subspace of dimension 1 in $J(M)$, that is, is a Lie subgroup $T \subset J(M)$ of dimension 1 ; and therefore one of the irreducible components of $W_{g-1} \cap (W_{g-1} + u)$ must be of the form $T+x$ for some point $x \in J(M)$, as desired.

(iii) Finally in this the most interesting case the divisors in $\varphi^{-1}(k+u) \subset M^{(4)}$ can be written in the form $p_1 + p_1^1 + p_1^2 + p_1^3$, where $(p_1, p_1^i) \in V_i$ and the points p_1^i are uniquely determined by p_1 whenever $\varphi(p_1) \notin W_1 \cap (W_1 + u)$; and yet again the points p_1^i vary over M as p_1 does, so that for all but finitely many points $p_1 \in M$ it follows that $\varphi(p_1^i) \notin W_1 \cap (W_1 + u)$ for any i as well. As in the preceding case, it is possible to introduce the group of admissible permutations for this configuration, which is clearly a transitive subgroup of order 4 and index 6 in the permutation group on 4 letters, and is thus isomorphic either to $\mathbf{Z}_4$ or to $\mathbf{Z}_2 + \mathbf{Z}_2$.

If the group of admissible permutations is isomorphic to $\mathbf{Z}_4$ it can be assumed by suitable labeling that this group is generated by the permutation of order 4 taking the ordered divisor $p_1 + p_1^1 + p_1^2 + p_1^3$ to the ordered divisor $p_1^1 + p_1^2 + p_1^3 + p_1$. The irreducible component $\theta(V_1)$ of the intersection $W_{g-1} \cap (W_{g-1} + u)$ is the image of the analytic mapping $p_1 \longrightarrow \varphi(p_1 + p_1^1) = (I + C_1)\varphi(p_1) + c_1$, using equations (1) and (2) again, so that $\theta(V_1) = (I + C_1)W_1 + c_1$; and similarly $\theta(V_2) = (I + C_2)W_1 + c_2$ and $\theta(V_3) = (I + C_3)W_1 + c_3$. On the other hand considering the permuted divisor $p_1^1 + p_1^2 + p_1^3 + p_1$ it also follows that $\varphi(p_1^1 + p_1) \in \theta(V_3)$, and hence in view of the irreducibility of these subvarieties that $(C_1 + I)W_1 + c_1 = \theta(V_3)$; thus $\theta(V_1) = \theta(V_3)$, so the intersection $W_{g-1} \cap (W_{g-1} + u)$ still has at most two components. Now applying equations (1) and (2) to these two divisors note that $\varphi(p_1^1) = C_1\varphi(p_1) + c_1$, $\varphi(p_1^2) = C_2\varphi(p_1) + c_2 = C_1[C_1\varphi(p_1) + c_1] + c_1$, $\varphi(p_1^3) = C_3\varphi(p_1) + c_3 = C_2[C_1\varphi(p_1) + c_1] + c_2$, and $\varphi(p_1) = C_3[C_1\varphi(p_1) + c_1] + c_3$; and since the points $\varphi(p_1)$ can lie in no proper linear subspace of $J(M)$ it follows immediately from this that $C_1^2 = C_2$, $C_1^3 = C_3$, and $C_1^4 = I$. The eigenvalues of the matrix C_1 are thus fourth roots of unity; and since $0 = I + C_1 + C_2 + C_3 = I + C_1 + C_1^2 + C_1^3$ as a consequence of (4), none of these eigenvalues can be 1. Not all the eigenvalues of C_1 can be -1, since then $I+C_1=0$. If two eigenvalues of C_1 are -1 the matrix $I+C_1$ is of rank 1, and consequently $\theta(V_1)$ must be a linear subspace of dimension 1 in $J(M)$; while if fewer than two eigenvalues of C_1 are -1 then at least two of these eigenvalues are $\pm i$ and the matrix $I+C_2=I+C_1^2$ is of rank 1, and consequently $\theta(V_2)$ must be a linear subspace of dimension 1 in $J(M)$. Thus at least one of the irreducible components of $W_{g-1}\cap(W_{g-1}+u)$ must be of the form $T+x$ where $T\subset J(M)$ is a Lie subgroup of dimension 1 and x is some point of $J(M)$, as desired.

If the group of admissible permutations is isomorphic to $\mathbf{Z}_2+\mathbf{Z}_2$ it can be assumed by suitable labeling that this group is generated by the two permutations of order 2 taking the ordered divisor $p_1+p_1^1+p_1^2+p_1^3$ to the ordered divisors $p_1^1+p_1+p_1^3+p_1^2$ and $p_1^2+p_1^3+p_1+p_1^1$ respectively. Applying equations (1) and (2) to these various divisors and comparing terms as in the above argument, it is easy to see that the linear transformations C_i must be such that

(11) $$C_1^2=C_2^2=C_3^2=I\,,\qquad C_1C_2=C_2C_1=C_3\,;$$

of course this merely amounts to the observation that these matrices form a representation of the group $\mathbf{Z}_2+\mathbf{Z}_2$. It follows from (4) that these matrices must also be such that

(12) $$I+C_1+C_2+C_3=0\,.$$

Now it is an easy consequence of (11) and (12) that either these matrices can be reduced to the standard forms

$$(13)\qquad C_1 = \begin{pmatrix} 1 & 0 & 0 \\ 0 & -1 & 0 \\ 0 & 0 & -1 \end{pmatrix}, \quad C_2 = \begin{pmatrix} -1 & 0 & 0 \\ 0 & 1 & 0 \\ 0 & 0 & -1 \end{pmatrix}, \quad C_3 = \begin{pmatrix} -1 & 0 & 0 \\ 0 & -1 & 0 \\ 0 & 0 & 1 \end{pmatrix}$$

or at least one of these matrices must be $-I$. On the other hand the irreducible component $\theta(V_i)$ of the intersection $W_{g-1} \cap (W_{g-1} + u)$ is the image of the analytic mapping $p_1 \longrightarrow \varphi(p_1 + p_1^i) = (I + C_i)\varphi(p_1) + c_i$, so that $\theta(V_i) = (I + C_i)W_1 + c_i$. Therefore it is impossible that $C_i = -I$ for any index i; and hence it follows from (13) that the matrices $I + C_i$ are all of rank one with linearly independent images. The irreducible components $\theta(V_i)$ are consequently all of the form $T_i + x_i$ for Lie subgroups $T_i \subset J(M)$ of dimension 1 and points $x_i \in J(M)$, and are moreover all distinct. That then suffices to complete the proof of the Corollary.

<u>Corollary 3 to Theorem 18.</u> For a Riemann surface of genus $g = 4$ and any point $u \neq 0$ in $J(M)$, the intersection $W_{g-1} \cap (W_{g-1} + u)$ has at most 2 irreducible components; and if it has precisely 2 irreducible components, then either $u \in W_1 - W_1$ and these components are as described in Theorem 17(b), or one of these components is of the form $X + T$ where X is an irreducible subvariety of dimension 1 and T is a Lie subgroup of dimension 1 in $J(M)$, or there is an endomorphism $C: J(M) \longrightarrow J(M)$ of period 3 with no real eigenvalues and these components are of the form

$C_1(W_1) - C_1^2(W_1) + c_1$ and $C_1^2(W_1) - C_1(W_1) + c_2$ for some points c_1, c_2 in $J(M)$.

Proof. When $g = 4$ then in addition to the regular cases discussed in the proof of Theorem 18 there occurs one exceptional case, in which the values of the parameters associated to the two standard irreducible components of the analytic subvariety $V \subset M^{(g-2)} \times M$ are

$$d_1 = 2, \quad r_1 = 4, \quad \nu_1 = 2 ,$$
$$d_2 = 2, \quad r_2 = 4, \quad \nu_2 = 2 .$$

With the notation as in the proof of Theorem 18, it follows as usual that the divisors in $\varphi^{-1}(k+u) \subset M^{(2g-2)} = M^{(6)}$ can be written in the form $(p_1 + p_2) + (p_1^1 + p_2^1) + (p_1^2 + p_2^2)$, where $(p_1 + p_2, p_j^j) \in V_i$ and the points p_j^i are determined uniquely (up to the interchange of p_1^i and p_2^i for each i) by the divisor $\mathfrak{d} = p_1 + p_2$ whenever $\varphi(\mathfrak{d}) \notin W_2 \cap (W_2 + u)$; and as the divisor $\mathfrak{d}$ varies over $M^{(2)}$ so do each of the divisors $\mathfrak{d}_i = p_1^i + p_2^i$, since $d_i = 2$. Then so long as the divisor $\mathfrak{d}$ lies outside a suitable proper analytic subvariety of $M^{(2)}$ it will also be the case that $\varphi(\mathfrak{d}_i) \notin W_2 \cap (W_2 + u)$ for $i = 1,2$; hence the order of the points in one of these divisors can be altered by a suitable permutation so that the divisor is rewritten in the form $(q_1 + q_2) + (q_1^1 + q_2^1) + (q_1^2 + q_2^2)$, where again $(q_1 + q_2, q_j^i) \in V_i$ and $\varphi(q_1 + q_2) \notin W_2 \cap (W_2 + u)$ but where either $q_1 + q_2 = \mathfrak{d}_1$ or $q_1 + q_2 = \mathfrak{d}_2$. Recalling that the divisor $\mathfrak{d}$ can be moved along a closed loop in $M^{(2)}$ avoiding the subset $\varphi^{-1}(W_2 \cap (W_2 + u)) \subset M^{(2)}$ in such a manner the natural analytic continuation of these divisors will interchange the points p_1^i and p_2^i, it is clear that $q_1^i + q_2^i$ will also be one of the divisors $\mathfrak{d}$, $\mathfrak{d}_1$, $\mathfrak{d}_2$; there thus arises a group of admissible permutations of the divisors $\mathfrak{d} + \mathfrak{d}_1 + \mathfrak{d}_2$ preserving the separate divisors $\mathfrak{d}$, $\mathfrak{d}_1$, and $\mathfrak{d}_2$. After suitable relabeling if necessary, this group can be taken to be the cyclic group of order 3 generated by the permutation taking the ordered divisor $\mathfrak{d} + \mathfrak{d}_1 + \mathfrak{d}_2$ to the order divisor $\mathfrak{d}_1 + \mathfrak{d}_2 + \mathfrak{d}$. Now applying equations (1) and (2) to these divisors it follows that $\varphi(\mathfrak{d}_2) = C_2\varphi(\mathfrak{d}) + c_2 = C_1[C_1\varphi(\mathfrak{d}) + c_1] + c_1$ and $\varphi(\mathfrak{d}) = C_2[C_1\varphi(\mathfrak{d}) + c_1] + c_2$; and since the points $\varphi(\mathfrak{d})$ can be contained in no proper linear subspace of $J(M)$ necessarily $C_1^2 = C_2$ and $C_1^3 = I$. Note that none of the eigenvalues of the matrix C_1 can be 1, since $0 = I + C_1 + C_2 = I + C_1 + C_1^2$ as a consequence of equation (4). Considering the divisor $\mathfrak{d}_1 + \mathfrak{d}_2 + \mathfrak{d}$ it is apparent that the irreducible component $\theta(V_2)$ of the intersection $W_{g-1} \cap (W_{g-1} + u)$ is the image of the natural analytic continuation of the mapping $\mathfrak{d} \longrightarrow \varphi(\mathfrak{d}_1 + p_1) = C_1\varphi(p_1 + p_2) + \varphi(p_1) + c_1 =$ $= C_1\varphi(p_2) - C_2\varphi(p_1) + c_1$, hence $\theta(V_2) = C_1(W_1) - C_1^2(W_1) + c_1$; and similarly, considering the divisor $\mathfrak{d}_2 + \mathfrak{d} + \mathfrak{d}_1$, it follows that $\theta(V_1) = C_1^2(W_1) - C_1(W_1) + c_2$. That then suffices to complete the proof of the Corollary.

(d) It has already been observed that the subvariety $W_1 \subset J(M)$ is fully determined by the two subvarieties W_{g-1} and W_{g-2}, since

$$W_1 = W_{g-1} \ominus W_{g-2} = \bigcap_{u \in W_{g-2}} (W_{g-1} - u) ;$$

and since W_1 is analytically equivalent to the Riemann surface M itself, it follows immediately that if two compact Riemann surfaces determine the same Jacobi variety and subvarieties W_{g-1} and W_{g-2} then they are necessarily analytically equivalent surfaces. Actually a considerably stronger result is an easy consequence of the preceding discussion of intersection properties in Jacobi varieties.

<u>Theorem 19. (Torelli's Theorem)</u> If M and M' are two compact Riemann surfaces of the same genus $g \geqq 5$ with Jacobi varieties $J(M)$ and $J(M')$ and subvarieties of positive divisors $W_r \subseteq J(M)$ and $W_r' \subseteq J(M')$ respectively, and if there is an analytic homeomorphism between the manifolds $J(M)$ and $J(M')$ transforming the subvariety $W_{g-1} \subseteq J(M)$ to the subvariety $W_{g-1}' \subseteq J(M')$, then the Riemann surfaces M and M' are also analytically homeomorphic.

Proof. To prove the desired result it is of course sufficient to show that the Riemann surface M itself can be reconstructed from knowledge only of the complex torus $J(M)$ and of the subvariety $W_{g-1} \subseteq J(M)$. For this purpose consider those points $u \in J(M)$ such that the intersection $W_{g-1} \cap (W_{g-1} + u)$ has precisely 2 irreducible components; it follows from Theorem 18 that there exist such points u , and that for any such point u either the 2 irreducible components are of the form $W_{g-2}+v_1$ and $-W_{g-2}+v_2$ for some points $v_1, v_2 \in J(M)$, or one of the components is of the form $X+T$ where $X \subseteq J(M)$ is an irreducible subvariety of dimension $g-3$ and $T \subseteq J(M)$ is a Lie subgroup of dimension 1. These two cases are readily distinguished, for a subvariety of the form $X+T$ is translated into itself by adding to it any one of the infinitely many points of the subgroup $T \subseteq J(M)$ while a subvariety of the form $W_{g-2}+v_1$ or $-W_{g-2}+v_2$ cannot be translated into itself by any nonzero point of $J(M)$, as noted on page 45; thus it is possible to construct a subvariety $V \subseteq J(M)$ such that either $V = W_{g-2}+v_1$ or $V = -W_{g-2}+v_2$ for some points $v_1, v_2 \in J(M)$. Now note that $W_{g-1} \ominus (W_{g-2}+v_1) = W_1 - v_1$ and that $W_{g-1} \ominus (-W_{g-2}+v_2) = (k-W_{g-1}) \ominus (-W_{g-2}+v_2) = -[(W_{g-1}-k) \ominus (W_{g-2}-v_2)]$ $= -W_1 + k - v_2$; and consequently either $W_{g-1} \ominus V = W_1 - v_1$ or $W_{g-1} \ominus V = -W_1 + k - v_2$. In any case the analytic subvariety $W_{g-1} \ominus V$ is analytically homeomorphic to the Riemann surface M itself; and since the construction involved only the complex torus $J(M)$ and the subvariety $W_{g-1} \subseteq J(M)$, the proof of the theorem is completed.

<u>Corollary 1 to Theorem 19.</u> The theorem also holds as stated for compact Riemann surfaces of genus $g \leqq 4$, provided that in the case $g = 4$ the Jacobi variety $J(M)$ does not admit an endomorphism of period 3 with no real eigenvalues.

Proof. There is only one compact Riemann surface of genus $g = 0$, while a surface of genus $g = 1$ is analytically homeomorphic to its Jacobi variety and a surface of genus $g = 2$ is analytically homeomorphic to the subvariety $W_{g-1} \subseteq J(M)$; the theorem is thus completely trivial in these cases. That the proof of theorem 19 goes through for compact Riemann surfaces of genus $g = 3$ and $g = 4$, with the exception as noted, follows immediately from Corollaries 2 and 3 to Theorem 18, respectively. That suffices for the proof of the desired result.

The difficulty in carrying the proof through in the exceptional case when $g = 4$ is merely that of intrinsically distinguishing between subvarieties of the form $\pm (W_2 + v)$ and those of the form $\pm (C_1W_1 - C_1^2W_1 + v)$, where C_1 is an endomorphism of $J(M)$ of period 3 with no real eigenvalues; and the difficulty lies in the method of proof rather than the theorem itself, which holds in this case as well. Rather than giving a separate argument in this special case, though, the reader will be referred to the other proofs of the theorem listed in the references, or left to conclude the proof on his own.

As interesting as it may be on its own, this theorem gains immeasurably greater significance if one is also aware that the subvariety $W_{g-1} \subset J(M)$ can be constructed quite explicitly, at least up to a translation in $J(M)$, from the period matrix of the marked Riemann surface alone; hence the Torelli theorem really implies that two marked Riemann surfaces are analytically equivalent if they have the same period matrices. These period matrices are an interesting set of moduli for describing Riemann surfaces, and their properties have not yet been altogether sorted out. It should be remarked that the period matrix of a marked Riemann surface determines not only just the Jacobi variety of that surface, but also the additional structure embodied in the naturally associated Riemann matrix pair; and this additional structure, sometimes called a polarization of the complex torus, is essential in describing the hypersurface W_{g-1}. Alternatively it is sometimes the pair consisting of the complex torus $J(M)$ and the analytic subvariety $W_{g-1} \subset J(M)$ that is called a polarized complex torus. The continuation of this tale must be left for another episode in the serial, though.

Notes for §4.

(a) It is a familiar result in the theory of functions of several complex variables that for any nonempty irreducible component V of the intersection of two analytic subvarieties of dimension r in a complex manifold of dimension g, necessarily $\dim V \geqq 2r-g$; hence an immediate consequence of Theorem 17(a) is that for any nonempty irreducible component V of the analytic subvariety W^2_{r+1} for any $r < g$, necessarily $\dim V \geqq 2r-g$. This result is of course just the special case $\nu = 2$ of Theorem 14(b). Actually a rather straightforward extension of Theorem 17(a) can be used to give another proof of the general case of Theorem 14(b) as well; this proof is due to H. Martens, and can be found in [4].

(b) The various properties of complex analytic subvarieties used more freely in this section are treated in most of the standard texts on functions of several complex variables. It should perhaps particularly be noted that subvarieties of codimension one in a complex manifold are characterized by the property that they are locally the sets of zeros of a single holomorphic function.

Intersections of the form $W_r \cap (W_{g-1} + u)$ can be described in a quite similar manner for any $r = 1,\ldots,g-1$; but for $r < g-1$ there are many more cases to consider than in the proof of Theorem 18, and it seems rather doubtful that the results are worth the effort involved in disentangling all the possibilities that arise.

A comment on the notation is perhaps in order here. It seems pointless to maintain separate notations for the Jacobi homomorphism $\Gamma(M, \mathcal{D}) \longrightarrow J(M)$ and the Jacobi mappings $M^r \longrightarrow J(M)$ and $M^{(r)} \longrightarrow J(M)$, once the properties of these various mappings have been established; so contrary to the practice adopted in §3 the same letter φ has been used for all of these mappings, leaving them to be distinguished either by context or explicitly as necessary.

(c) The existence of endomorphisms $C: J(M) \longrightarrow J(M)$ other than the trivial ones defined by matrices of the form nI for arbitrary integers n imposes rather severe restrictions on the complex torus $J(M)$; thus the exceptional cases considered in this section, especially that described in Corollary 3 to Theorem 18, can only occur for Riemann surfaces whose Jacobi varieties are of quite special forms. It would be of some interest to see whether these cases can indeed occur at all. The endomorphisms of non-trivial sort are called complex multiplications of the torus $J(M)$; their investigation is an interesting subject in its own right, and has been the subject of an extensive literature. (See [26].)

(d) The proof of Torelli's theorem given here, based on the analysis of intersections in Jacobi varieties in the preceding sections (b) and (c), is essentially that given by A. Weil in [10], translated from the algebro-geometric to the analytic point of view; separate proofs of Torelli's theorem in the cases of genus

$g = 3$ and $g = 4$, avoiding the complications discussed in section (c) and the remaining gap in the proof in these notes, can also be found in [10]. A rather shorter and more direct proof of Torelli's theorem, based on an extension of the techniques discussed in section (a) rather than on the general discussion of intersections in section (b), was given by H. H. Martens in [2] and [3]. Other proofs can be found in the references listed next.

References. The main source for the material discussed in this chapter is [10]. Other treatments of Torelli's theorem and discussions of related results can be found in the following, and in [2], [3], [7].

[23] Andreotti, Aldo, Recherches sur les surfaces algébriques irregulières, Mém. Acad. Belg. 27(1952), fasc. 7.

[24] -----, On a theorem of Torelli, Amer. Jour. of Math. 80 (1958), 801-828.

[25] Torelli, R., Sulle varietà di Jacobi, Rend. Accademia Lincei 22(1914), 98-103.

[26] Weil, André, On the theory of complex multiplication, Proc. International Symposium on Algebraic Number Theory, Tokyo (1955), 9-22.

Appendix. On conditions ensuring that $W_r^2 \neq \emptyset$.

One special case of the topological argument outlined on page 104 is of particular interest; and since it involves only quite well known topological techniques in a very straightforward manner, it is perhaps worthwhile appending a more detailed discussion of that case.

If for some index r the subvariety $G_r^2 \subseteq M^{(r)}$ is empty, then the Jacobi mapping $\varphi: M^{(r)} \longrightarrow J(M)$ is a nonsingular complex analytic homeomorphism between $M^{(r)}$ and the complex analytic submanifold $W_r \subseteq J(M)$; thus $M^{(r)}$ can be viewed as a regularly imbedded analytic submanifold of $J(M)$. As such there is a well defined analytic normal bundle $N(M^{(r)})$, which is a complex analytic vector bundle of rank $g-r$ over the manifold $M^{(r)}$; and the direct sum of the tangent bundle $T(M^{(r)})$ and this normal bundle $N(M^{(r)})$ is topologically equivalent to the restriction to the submanifold $M^{(r)} \subseteq J(M)$ of the tangent bundle to $J(M)$, hence that direct sum is topologically equivalent to a trivial vector bundle of rank g over the manifold $M^{(r)}$. Therefore introducing the total Chern classes

$$c(T(M^{(r)})) = 1 + c_1 + \dots + c_r , \qquad c(N(M^{(r)})) = 1 + n_1 + \dots + n_{g-r}$$

where $c_i, n_i \in H^{2i}(M,\mathbf{Z})$, it follows that

$$(1) \qquad c(T(M^{(r)}))\cdot c(N(M^{(r)})) = 1 .$$

The topological properties of the symmetric products $M^{(r)}$ have been described very conveniently by I. G. Macdonald in [18], and can be summarized briefly as follows. The cohomology ring $H^*(M^{(r)},\mathbb{Z})$ is generated by elements $\xi_1,\ldots,\xi_g$, $\xi_1',\ldots,\xi_g'$ of degree 1 and an element η of degree 2; these are subject to the usual commutativity relations, namely that elements of degree 1 anticommute with one another and commute with elements of degree 2, and in addition to the relations

$$(2) \qquad \xi_{i_1} \cdots \xi_{i_a} \xi_{j_1}' \cdots \xi_{j_b}' (\xi_{k_1}\xi_{k_1}' - \eta) \cdots (\xi_{k_c}\xi_{k_c}' - \eta)\eta^q = 0$$

where $i_1,\ldots,i_a$, $j_1,\ldots,j_b$, $k_1,\ldots,k_c$ are any distinct integers and a, b, c, q are any integers such that $a+b+2c+q = r+1$. In particular $H^{2r}(M^{(r)},\mathbb{Z})$ is generated by the single element η^r, and $H^{2r-2}(M^{(r)},\mathbb{Z})$ is generated by the $\binom{2g}{2}+1$ independent elements η^{r-1}, $\xi_{i_1}\xi_{i_2}\eta^{r-2}$, $\xi_{i_1}\xi_{j_1}'\eta^{r-2}$, $\xi_{j_1}'\xi_{j_2}'\eta^{r-2}$ where $i_1 < i_2$; $j_1 < j_2$; as usual $\binom{2g}{2}$ denotes a binomial coefficient. Setting $\sigma_i = \xi_i\xi_i' \in H^2(M^{(r)},\mathbb{Z})$ and $\varphi_i = \sigma_i - \eta \in H^2(M^{(r)},\mathbb{Z})$, the total Chern class of the tangent bundle to the manifold $M^{(r)}$ is

$$(3) \qquad c(T(M^{(r)})) = (1+\eta)^{r-2g+1} \prod_{i=1}^{g} (1-\varphi_i) .$$

These assertions are demonstrated in [18], with the same notation. An additional useful observation is the following.

<u>Lemma 8.</u> For any index $0 \leqq a \leqq r$ and any distinct integers $i_1,\ldots,i_a$ in $[1,g]$,

$$\sigma_{i_1} \cdots \sigma_{i_a} \eta^{r-a} = \eta^r ;$$

and for any index $0 \leqq a \leqq r-1$ and any distinct integers $i_1,\ldots,i_a$

in $[1,g]$,

$$\sigma_{i_1} \cdots \sigma_{i_a} \eta^{r-1-a} = (\sigma_{i_1} + \ldots + \sigma_{i_a})\eta^{r-2} + (1-a)\eta^{r-1} .$$

Proof. Both relations hold trivially for $a = 0$, and the second relation also holds trivially for $a = 1$; and will be demonstrated by induction on the index a . To prove the first relation note that as a consequence of (2) it follows that $\varphi_{i_1} \cdots \varphi_{i_a} \eta^{r+1-a} = 0$ whenever $1 \leqq a \leqq \frac{1}{2}(r+1)$; and evidently then $\varphi_{i_1} \cdots \varphi_{i_a} \eta^{r-a} = 0$ whenever $1 \leqq a \leqq r$. Assuming that the first relation holds for all indices less than a and considering the expansion of the product $(\sigma_{i_1} - \eta) \ldots (\sigma_{i_a} - \eta)\eta^{r-a} = 0$, all of the terms in the expansion of this product except for the term $\sigma_{i_1} \cdots \sigma_{i_a} \eta^{r-a}$ coincide with the corresponding terms in the expansion of the product $(\eta - \eta) \ldots (\eta - \eta)\eta^{r-a}$; and it follows immediately therefore that $\sigma_{i_1} \cdots \sigma_{i_a} \eta^{r-a} = \eta^r$ as desired. To prove the second relation note as above that as a consequence of (2) it follows that $\varphi_{i_1} \cdots \varphi_{i_a} \eta^{r-1-a} = 0$ whenever $1 \leqq a \leqq r$. Assuming then that the second relation holds for all indices less than a it follows that

$$0 = (\sigma_{i_1} - \eta) \cdots (\sigma_{i_a} - \eta)\eta^{r-1-a}$$

$$= \sum_{\nu=0}^{a} \sum_{1 \le j_1 < \dots < j_\nu \le a} (-1)^{a-\nu} \sigma_{i_{j_1}} \cdots \sigma_{i_{j_\nu}} \eta^{r-1-\nu}$$

$$= \sigma_{i_1} \cdots \sigma_{i_a} \eta^{r-1-a} +$$

$$+ \sum_{\nu=0}^{a-1} \sum_{1 \le j_1 < \dots < j_\nu \le a} (-1)^{a-\nu} [(\sigma_{i_{j_1}} + \dots + \sigma_{i_{j_\nu}})\eta^{r-2} + (1-\nu)\eta^{r-1}]$$

$$= \sigma_{i_1} \cdots \sigma_{i_a} \eta^{r-1-a} +$$

$$+ \sum_{\nu=1}^{a-1} (-1)^{a-\nu} \binom{a-1}{\nu-1} (\sigma_{i_1} + \dots + \sigma_{i_a})\eta^{r-2} + \sum_{\nu=0}^{a-1} (-1)^{a-\nu} \binom{a}{\nu} (1-\nu)\eta^{r-1}$$

$$= \sigma_{i_1} \cdots \sigma_{i_a} \eta^{r-1-a} - (\sigma_{i_1} + \dots + \sigma_{i_a})\eta^{r-2} - (1-a)\eta^{r-1} \text{ , when } a > 1 \text{ ,}$$

which yields the desired result and concludes the proof of the lemma.

Combining the preceding observations leads rather directly to the following result.

Lemma 9. If for some index r the subvariety $G_r^2 \subseteq M^{(r)}$ is empty, then the Chern classes of the normal bundle $N(M^{(r)})$ have the form

$$n_r = \sum_{\mu=0}^{g} (-1)^{r-\mu} \binom{g}{\mu} \binom{2r-g}{r-\mu} \cdot \eta^r$$

and

$$n_{r-1} = \sum_{\mu=0}^{g} (-1)^{r-1-\mu} (1-\mu) \binom{g}{\mu} \binom{2r-1-g}{r-1-\mu} \cdot \eta^{r-1}$$

$$+ \sum_{\mu=1}^{g} (-1)^{r-1-\mu} \binom{g-1}{\mu-1} \binom{2r-1-g}{r-1-\mu} \cdot (\sigma_1 + \dots + \sigma_g)\eta^{r-2} \text{ ,}$$

with the usual notation for the binomial coefficients.

Proof. If the subvariety $G_r^2 \subseteq M^{(r)}$ is empty then it follows from (1) and (3) that

$$
\begin{aligned}
c(N(M^{(r)})) &= (1+\eta)^{2g-r-1} \prod_{i=1}^{g} (1-\varphi_i)^{-1} \\
&= \sum_{\nu,\nu_1,\ldots,\nu_g \geqq 0} \binom{2g-r-1}{\nu} \cdot \eta^{\nu} \varphi_1^{\nu_1} \cdots \varphi_g^{\nu_g} \\
&= \sum_{\nu,\nu_1,\ldots,\nu_g \geqq 0} (-1)^{\nu_1+\ldots+\nu_g} \binom{2g-r-1}{\nu} \cdot \\
&\qquad \cdot \eta^{\nu} (\eta^{\nu_1} - \nu_1 \eta^{\nu_1-1} \sigma_1) \cdots (\eta^{\nu_g} - \nu_g \eta^{\nu_g-1} \sigma_g) ,
\end{aligned}
$$

recalling that $\varphi_i = \sigma_i - \eta$ and that $\sigma_i^2 = 0$. The class $n_r \in H^{2r}(M^{(r)},\mathbf{Z})$ consists of those terms of degree $2r$ in the above summation; and applying the first relation in Lemma 8 and expanding the summation, it follows that

$$
\begin{aligned}
n_r &= \sum_{\substack{\nu+\nu_1+\ldots+\nu_g=r \\ \nu \geqq 0, \quad \nu_i \geqq 0}} (-1)^{\nu_1+\ldots+\nu_g} \binom{2g-r-1}{\nu} (1-\nu_1) \cdots (1-\nu_g) \cdot \eta^r \\
&= \sum_{\nu+\nu_1+\ldots+\nu_g=r} \sum_{\mu=0}^{g} \sum_{1 \leqq j_1 < \cdots < j_\mu \leqq g} (-1)^{r-\nu-\mu} \binom{2g-r-1}{\nu} \nu_{j_1} \cdots \nu_{j_\mu} \cdot \eta^r \\
&= \sum_{\nu+\nu_1+\ldots+\nu_g=r} \sum_{\mu=0}^{g} (-1)^{r-\nu-\mu} \binom{2g-r-1}{\nu} \binom{g}{\mu} \nu_1 \cdots \nu_\mu \cdot \eta^r .
\end{aligned}
$$

Note that

$$(4) \qquad \sum_{\substack{\nu_1+\cdots+\nu_g=r \\ \nu_i \geqq 0}} \nu_1 \cdots \nu_\mu = \binom{g-1+r}{g-1+\mu}$$

for any indices $0 \leq \mu \leq g$, $0 \leq r$; if not well known, this is at least easily demonstrated by induction, and the details will be omitted. Applying (4) it then follows that

$$n_r = \sum_{\mu=0}^{g} \sum_{\nu=0}^{r} (-1)^{r-\nu-\mu} \binom{2g-r-1}{\nu} \binom{g}{\mu} \binom{g-1+r-\nu}{g-1+\mu} \cdot \eta^r ;$$

and this can be simplified by applying a standard form of the Vandermonde convolution formula [see J. Riordan, Combinatorial Identities (Wiley, New York, 1968), formula (5) on page 8], yielding the result that

$$n_r = \sum_{\mu=0}^{g} (-1)^{r-\mu} \binom{g}{\mu} \binom{2r-g}{r-\mu} \cdot \eta^r ,$$

as desired. The class $n_{r-1} \in H^{2r-2}(M^{(r)}, \mathbb{Z})$ consists of those terms of degree $2r-2$ in the original summation; and expanding the resultant summation and applying the second relation in Lemma 8, it follows that

$$n_{r-1} = \sum_{\substack{\nu+\nu_1+\ldots+\nu_g=r-1 \\ \nu \geq o, \quad \nu_i \geq o}} (-1)^{\nu_1+\ldots+\nu_g} \binom{2g-r-1}{\nu} \eta^{\nu} (\eta^{\nu_1} - \nu_1 \eta^{\nu_1-1} \sigma_1) \ldots (\eta^{\nu_g} - \nu_g \eta^{\nu_g-1} \sigma_g)$$

$$= \sum_{\nu+\nu_1+\ldots+\nu_g=r-1} (-1)^{r-1-\nu} \binom{2g-r-1}{\nu} \cdot \eta^{r-1}$$

$$+ \sum_{\nu+\nu_1+\ldots+\nu_g=r-1} \sum_{\mu=1}^{g} \sum_{1 \leq j_1 < \ldots < j_\mu \leq g} (-1)^{r-1-\nu-\mu} \binom{2g-r-1}{\nu} \nu_{j_1} \ldots \nu_{j_\mu} \cdot \eta^{r-1-\mu} \sigma_{j_1} \ldots \sigma_{j_\mu}$$

$$= \sum_{\nu+\nu_1+\ldots+\nu_g=r-1} (-1)^{r-1-\nu} \binom{2g-r-1}{\nu} \cdot \eta^{r-1}$$

$$+ \sum_{\nu+\nu_1+\ldots+\nu_g=r-1} \sum_{\mu=1}^{g} \sum_{1 \leq j_1 < \ldots < j_\mu \leq g} (-1)^{r-1-\nu-\mu} \binom{2g-r-1}{\nu} \nu_{j_1} \ldots \nu_{j_\mu} \cdot [(\sigma_{j_1} + \ldots + \sigma_{j_\mu}) \eta^{r-2} + (1-\mu) \tau$$

$$= \sum_{\nu+\nu_1+\ldots+\nu_g=r-1} \sum_{\mu=o}^{g} (-1)^{r-1-\nu-\mu} \binom{2g-r-1}{\nu} \binom{g}{\mu} (1-\mu) \nu_1 \ldots \nu_\mu \cdot \eta^{r-1}$$

$$+ \sum_{\nu+\nu_1+\ldots+\nu_g=r-1} \sum_{\mu=1}^{g} (-1)^{r-1-\nu-\mu} \binom{2g-r-1}{\nu} \binom{g-1}{\mu-1} \nu_1 \ldots \nu_\mu \cdot (\sigma_1 + \ldots + \sigma_g) \eta^{r-2} .$$

Applying (4) once more this can be rewritten

$$n_{r-1} = \sum_{\mu=0}^{g} \sum_{\nu=0}^{r-1} (-1)^{r-1-\nu-\mu} \binom{2g-r-1}{\nu}\binom{g}{\mu}\binom{g-2+r-\nu}{g-1+\mu}(1-\mu)\cdot\eta^{r-1}$$

$$+ \sum_{\mu=1}^{g} \sum_{\nu=0}^{r-1} (-1)^{r-1-\nu-\mu} \binom{2g-r-1}{\nu}\binom{g-1}{\mu-1}\binom{g-2+r-\nu}{g-1+\mu}\cdot(\sigma_1 + \dots + \sigma_g)\eta^{r-2}$$

and after further simplification by an application of the Vandermonde convolution formula as before it follows that

$$n_{r-1} = \sum_{\mu=0}^{g} (-1)^{r-1-\mu} \binom{g}{\mu}\binom{2r-g-1}{r-1-\mu}(1-\mu)\cdot\eta^{r-1}$$

$$+ \sum_{\mu=1}^{g} (-1)^{r-1-\mu} \binom{g-1}{\mu-1}\binom{2r-g-1}{r-1-\mu}\cdot(\sigma_1 + \dots + \sigma_g)\eta^{r-2} ,$$

which serves to complete the proof of the lemma.

The formulas of Lemma 9 are already simple enough for the present purposes, even though further simplifications are still possible. Thus for the special case that $g = 2r-1$ note that $\binom{2r-g}{r-\mu} = \binom{1}{r-\mu}$ is nonzero only for $\mu = r$ or $r-1$, so the summation in the first formula of Lemma 9 can be restricted to these two indices and the result easily calculated; and the calculations are equally simple for the special cases that $g = 2r-2$ and $g = 2r-3$, and for both the first and second formulas of Lemma 9. The relevant results are the following:

(5) if $g = 2r-1$ then $n_r = 0$;

(6) if $g = 2r-2$ then $n_r = -\frac{2}{r}\binom{2r-2}{r-1}\cdot\eta^r$;

(7) if $g = 2r-3$ then $n_r = 0$ and

$$n_{r-1} = 2\,\frac{r-4}{r}\,\binom{2r-3}{r-1}\cdot\eta^{r-1} - 2\,\frac{r-3}{r(r-1)}\binom{2r-4}{r-2}\cdot(\sigma_1 + \dots + \sigma_g)\eta^{r-2}.$$

These observations then lead almost immediately to the principal result of this appendix.

Theorem 20. On a Riemann surface of genus $g \geqq 2$ if $2r - (g+2) \geqq 0$ then necessarily $W_r^2 \neq \emptyset$ and $G_r^2 \neq \emptyset$.

Proof. It clearly suffices merely to show that $G_r^2 \neq \emptyset$ for the least value of the index r such that $2r - (g+2) \geqq 0$, hence for the index r such that $g = 2r-2$ when g is even and for the index r such that $g = 2r-3$ when g is odd. If $G_r^2 = \emptyset$ when $g = 2r-2$ then since $g-r = r-2$ necessarily $n_r = 0$, where as before $n_r \in H^{2r}(M^{(r)},\mathbb{Z})$ is a Chern class for the normal bundle $N(M^{(r)})$, a vector bundle of rank $g-r$ over the manifold $M^{(r)}$; but (6) shows that $n_r \neq 0$, a contradiction. If $G_r^2 = \emptyset$ when $g = 2r-3$ then since $g-r = r-3$ necessarily $n_r = 0$ and $n_{r-1} = 0$; but (7) shows that $n_{r-1} \neq 0$, a contradiction again, and the proof is thereby completed.

This result completes Theorem 14 in a natural manner, at least in the case that $\nu = 2$. For Theorem 14(b) implies that for any nonempty irreducible component V of the analytic subvariety $W_r^2 \subseteq J(M)$ necessarily $\dim V \geqq 2r - (g+2)$, recalling Corollary 1 to that theorem; and Theorem 20 shows that whenever $2r - (g+2) \geqq 0$ there necessarily exist some nonempty components of the subvariety $W_r^2 \subseteq J(M)$. Combining these results with Theorem 7 it follows that

(8) $$2r - (g+2) \leqq \dim W^2_r \leqq r-2 .$$

A superficial glance at Theorem 14(a) might lead one to expect a somewhat better result than that given by Theorem 20, until one recalls that G^2_r is generally fibred over W^2_r with fibre dimension 1 and hence that any irreducible component of G^2_r must be of dimension at least 1; indeed it is apparent from (5) that the argument in Theorem 20 fails for the case that $g = 2r-1$.

<u>Corollary 1 to Theorem 20.</u> Any compact Riemann surface of genus g admits a representation as a branched analytic covering of the Riemann sphere $\mathbb{P}^1$ of at most $[\frac{g+1}{2}]+1$ sheets.

Proof. The assertion is trivial for $g = 0,1$. If $g \geqq 2$ it follows from Theorem 20 that $W^2_r \neq \emptyset$ whenever $2r - (g+2) \geqq 0$, hence in particular that $W^2_r \neq \emptyset$ for $r = [\frac{g+1}{2}]+1$. Thus for that value of r there must exist a complex analytic line bundle ξ with $c(\xi) = r$ and $\gamma(\xi) \geq 2$; and the quotient of any two linearly independent holomorphic sections of ξ is a meromorphic function which represents the given Riemann surface as a branched analytic covering of the Riemann sphere of at most r sheets, thereby completing the proof of the Corollary.

This Corollary indicates the particular interest associated to the problem of determining whether $W^2_r \neq \emptyset$; other proofs of the Corollary have been given by T. Meis [24], G. Kempf [12], and S. Kleiman and D. Laksov [14], and rather incomplete proofs appeared much earlier in the literature. The analogous proof that $W_r^\nu \neq \emptyset$ whenever $r\nu - (\nu-1)(g+\nu) \geqq 0$ requires rather more topological machinery, along the lines indicated in [12] and [14].

Index of symbols

Index

Milton Keynes UK
Ingram Content Group UK Ltd.
UKHW021026140924
448309UK00006B/301

9 780691 619255